# International Conference on Full Scale Ship Performance 2021



Online
10 - 11 February 2021

ISBN: 978-1-7138-8297-8

**Printed from e-media with permission by:**

Curran Associates, Inc.
57 Morehouse Lane
Red Hook, NY 12571

**Some format issues inherent in the e-media version may also appear in this print version.**



The Royal Institution of Naval Architects
8-9 Northumberland Street
London, WC2N 5DA
United Kingdom

Phone: 020 7235 4622
Fax: 020 7259 5912

publications@rina.org.uk

**Additional copies of this publication are available from:**

Curran Associates, Inc.
57 Morehouse Lane
Red Hook, NY 12571 USA
Phone: 845-758-0400
Fax: 845-758-2633
Email: curran@proceedings.com
Web: www.proceedings.com

# International Conference

# Full Scale Ship Performance

10-11 February 2021, Online Conference

## PROCEEDINGS

THE ROYAL INSTITUTION OF NAVAL ARCHITECTS
8-9 Northumberland Street, London, WC2N 5DA, Telephone: +44 (0)20 7235 4622

The Royal Institution of
Naval Architects

INTERNATIONAL CONFERENCE

# Full Scale Ship Performance

10-11 February 2021, Online Conference



Papers published in these conference Proceedings are peer reviewed.

THE ROYAL INSTITUTION OF NAVAL ARCHITECTS
8-9 Northumberland Street
London WC2N 5DA

Telephone: 020 7235 4622
Fax: 020 7259 5912

**CONTENTS**

# OPERATIONAL AND ENERGY USE PROFILE CHARACTERISATION OF LNG CARRIERS WITH DFDE PROPULSION PLANTS

**J Bonello,** UCL Energy Institute, UK
**C González Gutiérrez,** University of Cantabria, Spain
**S Suárez de la Fuente,** UCL Energy Institute, UK

## SUMMARY

Traditionally, the LNG fleet has been dominated by ships powered by steam plants however, new Four-stroke Dual Fuel Diesel Electric (DFDE) or Tri-Fuel Diesel Electric (TFDE) propulsive plants offer fuel flexibility while operating at a higher thermal efficiency. This propulsion option has become a relevant alternative to steam-powered plants representing about 32% of the total LNG fleet in 2018. This study seeks to establish the operational and energy use profile of LNG carriers based on the analysis of high-frequency in-service data. It focuses on LNG carriers with DFDE propulsion systems using data collected between 2016 and 2018 from 11 LNG carriers – representing about 7% of the DFDE and TFDE powered LNG fleet. The continuous monitoring data covers many performance parameters allowing for a reliable analysis of operation and energy use while reducing uncertainty by measuring essential parameters rather than assuming or approximating.

## NOMENCLATURE

| | |
|---|---|
| $C_F$ | Carbon factor ($t_{CO2}\, t_{fuel}^{-1}$) |
| $LHV$ | Lower Heating Value (MJ $kg^{-1}$) |
| $M$ | Mass flow rate (kg $h^{-1}$) |
| $MCR$ | Maximum Continuous Rating (%) |
| $S$ | Result of EMA at any time (n/a) |
| $t$ | Time (h) |
| $T$ | Temperature (°C) |
| $v$ | Speed Over Ground (kn) |
| $\dot{W}$ | Power (kW) |
| $Y$ | Result of EMA at specific time (n/a) |
| $\varepsilon$ | $CO_2$ emissions (t) |
| $\alpha$ | Degree of weighting decrease in EMA method (n/a) |
| $\eta$ | Efficiency (%) |

## 1. INTRODUCTION

The current global energy demand is transitioning towards fuels with less environmental impact making LNG (Liquified Natural Gas) a very attractive candidate. The increase in demand has resulted in the transport of natural gas to gain importance [1], [2] and, subsequently, the LNG Carrier (LNGC) fleet has expanded significantly [3] into an important segment of maritime transport ([3] [4]). As of the end of 2018, the LNGC fleet consisted of 525 ships, with heterogeneous propulsion systems, sizes, cargo systems and capacity [5].

Traditionally, the LNGC fleet has been dominated by steam propulsion since the 1960's [6]. However, changing trade patterns from traditional long-term contracts to a shorter term spot market requires higher flexibility and better fuel efficiency over a range of operating conditions. A significant increase in the volume of LNG traded on the spot market has been registered with volume rising to 31% of all trade; a 50% jump from 2008 to 2018. [5].

Despite the advantages of steam plants such as low maintenance costs and simpler designs (especially boil-off gas management), the low thermal efficiency causing higher transport costs, made it natural to find propulsion alternatives. The development of more efficient containment systems leading to lower boil-off rates, not be sufficient for steam plants, requiring forced boil-off using other fuels. These factors promoted the appearance of new generation dual or tri-fuel diesel electric propulsion systems (DFDE or TFDE) [7], [8] which, for simplicity, will be referred to as DFDE in this paper. The relevance of the DFDE LNGC is evidenced by the aggressive uptake with 32% of the fleet having them onboard in 2018 [9].

DFDE propulsion has a higher thermal efficiency and energy density gaining more cargo space, it is easier to shut down and start-up the plant and provides increased redundancy with most layouts having four main engines and two propulsion shafts. DFDE is also the main propulsion system found on Floating Regasification Units (FSRU) accounting for over half of the global fleet in 2019 [10].

Dual Fuel (DF) layouts offer high flexibility in fuel use including residual fuel (RO), distillate fuel (DO) and Boil-Off Gas (BOG). In addition to the conventional fuel supply lines, these ships also need secondary fuel lines to supply a small amount of fuel known as pilot fuel is used for combustion when operating in gas [11] and distillate mode [12] to ensure nozzle cooling. The operation and performance of the DF engines has been extensively documented in literature [13]–[21].

Despite the growing number of vessels with this propulsion type, there has been very sparse literature that analyses LNGC performance in operation and actual fuel use in granular detail. This study seeks to establish the operational and energy profile of the LNGC fleet based on the analysis of high-frequency in-service monitoring data.

 

It will focus on LNG carriers with DFDE propulsion systems using data collected from 11 LNGCs representing about 12% of the DFDE-powered LNGC fleet. The Continuous Monitoring (CM) data covers several performance parameters at high-frequency and with high level of accuracy without human error, increasing transparency in the results. This transparency eliminates one of the barriers on the reliability of the analysis as mentioned by Poulsen and Johnson (2016) and Soner et al. (2018) while reducing uncertainty by measuring essential parameters rather than assuming or approximating. By doing this, it can provide statistics describing the performance of LNGC powered by DFDE as a foundation for further modelling work.

The paper will start with a brief overview of characteristics of LNG carriers with steam plants, this will be followed by method, operational modelling, operational and energy use profiles and conclusions.

## 2. METHOD

The analyse of high frequency in-service data, collected by different on-board sensors (Table 1), is coupled with operative assumptions allowing for analysis into ship operation and energy use. Moreover, the high levels of telemetry on board is greatly helpful for shipping companies to achieve the optimisation management and deployment of the fleet [24]. This work includes over 120,000 hourly observations between the end of 2016 and beginning of 2018 for 11 LNGC operating under medium-long Time Charter Contract (TCC). No GPS data was provided with the dataset due to commercial sensitivity which limited the extent of the analysis possible. The method can be divided in five main blocks as in Figure 1.

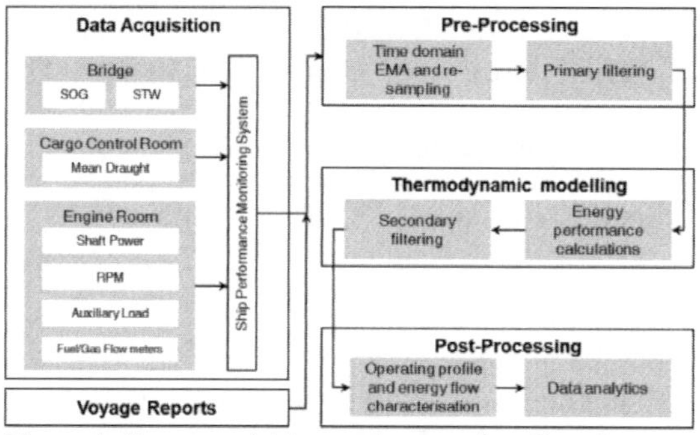


Figure 1: Data acquisition and processing flowchart

| Input | Instrumentation | Error Associated |
|---|---|---|
| Shaft Power | Torque-meter | 0.5 % |
| Shaft RPM | Torque-meter | 0.1 % |
| Fuel consumption | Volumetric flowmeters | 0.5-1.0 %[a] |
| Gas consumption | Coriolis flowmeters | 0.1-0.2 %[a] |
| Ship Speed Through Water (STW) | Electromagnetic & Doppler Log | 3.0 % [b] |
| Ship Speed Over Ground (SOG) | GPS | 5.0 % [b] |
| Mean Draught | Draught sensors | ±0.1 m [b] |
| DF Generators Power output | Kilowatt-meters | 0.1 % [c] |

[a][25]; [b][26]; [c][27]

Table 1: Instrumentation used for measuring the inputs on this study

### 2.1 DATA ACQUISITION

The data acquisition involves the recording and storage of data produced by the sensors connected to the Ship Performance Monitoring (SPM) system. The SPM receives data from the different systems onboard and averages data at 15s intervals using an Exponential Moving Average (EMA) between loggings (Equation 1).

$$S_t = (1 - \alpha) \cdot S_{t-1} + \alpha \cdot Y_t, \qquad t > 1 \qquad (1)$$

Where $S_t$ is the value of the EMA at any time period $t$, $\alpha$ represents the degree of weighting decrease and $Y_t$ is the value at time period $t$. Finally, the ship data is anonymised and saved into a CSV file.

### 2.2 PRE-PROCESSING

During the pre-processing stage, the data collected in CM is averaged for one-hour periods and manual primary filtering is applied to eliminate outliers (3.3% of the initial dataset) as it is shown in the Table 2. For certain voyages, it was not possible to clearly determine periods when the ship was under laden or ballast conditions which led the authors to eliminate observations for those voyages.

### 2.3 OPERATIONAL MODELLING

#### 2.3 (a) Power

The electrical and propulsion energy on board is produced by the DF engines connected to generators which distribute the electricity for the auxiliary load and propulsion. Vessels also have auxiliary boilers to support on-board heating demand.

| Input | Condition | Observations filtered | % of the total |
|---|---|---|---|
| Speed | \|SOG – STW\| > 4kn | 4206 | 3.3 |
| Mean Draught | Unstable loading condition | 17 | 0.0 |

Table 2: Observations eliminated during primary filtering

The cargo containment system allows heat transfer from the surroundings to both the liquid and the gaseous cargo phases, resulting in a continuous evaporation process of the LNG [28]. This vaporization process in the cargo tanks rise the internal tank pressure, to control the tank pressure this BOG is vented off. The BOG on LNG carriers with DFDE propulsion plants, can use the BOG as fuel for propulsion, generation of electric power or disposed of by sending it to the Gas Combustion Unit (GCU).

The plant's power input ($\dot{W}_{in}$) on DFDE LNG carriers is provided by the different fuels on board and the following approach is used to capture the different power sources:

$$\dot{W}_{in} = \frac{A+B+C+D}{3600} \tag{2}$$

$$A = M_{FO,i} \cdot LHV_{FO,i} \tag{3}$$

$$B = M_{BOG} \cdot LHV_{BOG} \tag{4}$$

$$C = M_{PO} \cdot LHV_{PO} \tag{5}$$

$$D = M_{FO-AB,i} \cdot LHV_{FO,i} \tag{6}$$

Where $M_{FO}$ and the $M_{FO-AB}$ are respectively the DF engines and auxiliary boiler hourly liquid fuel oil (FO) – RO or/and DO, $LHV_{FO}$ is the fuel's Lower Heating Value (LHV) according
Table 3, and *i* represents the FO being consumed (i.e. residual or distillate). The LHV has been used per the IMO Energy Efficiency Design Index Guidelines [29]. The $M_{PO}$ is the mass of pilot fuel used and the $LHV_{PO}$ is the lower heating value of the pilot fuel used, assuming the pilot fuel used is DO. The total Boil Of Gas (BOG) mass consumed in the DF engines is represented by $M_{BOG}$, $LHV_{BOG}$ is the LNG's LHV. It is important to highlight that $LHV_{BOG}$ has been assumed constant since the cargo general composition at the moment of loading and unloading was not recorded. The creation of a BOG composition model such as the one seen in Dimopoulos and Frangopoulos (2008) is out of the scope of this work.
Pilot fuel consumption is not measured but can be estimated based on operation mode (DO or BOG) and engine testbed data at different engine loads.

| | RO | DO | BOG |
|---|---|---|---|
| LHV (MJ $kg^{-1}$) | 40.2[a] | 42.7[a] | 48.0[c] |

[a][31], [c][32], [d][33]

Table 3: Calorific values assumed per fuel type

The power consumption ($\dot{W}_{out}$) on board the DFDE LNG carriers is captured by the following equation:

$$\dot{W}_{out} = \dot{W}_s + \dot{W}_{AL} + \dot{W}_{AB} \tag{7}$$

Where $\dot{W}_s$, $\dot{W}_{AL}$ ,and $\dot{W}_{AB}$ are the power take-off by the propeller shaft, auxiliary load, auxiliary boiler and gas combustion unit respectively and they were provided by the SPM. The amount of fuel power lost by the GCU unit ($\dot{W}_{GCU}$) can be quantified as follows:

$$\dot{W}_{GCU} = \frac{M_{BOG-GCU} \cdot LHV_{BOG}}{3600} \tag{8}$$

Where $M_{BOG-GCU}$ is the BOG consumed by the GCU provided by the SPM and the $LHV_{BOG}$ is the lower heating value of the LNG as given in
Table 3.

The power output for the auxiliary boiler is dependent of steam and feed water properties [34] and is estimated as follows:

$$\dot{W}_{AB} = \frac{(M_{FO-AB,i} \cdot LHV_{FO,i}) \cdot \eta_b}{3600} \tag{9}$$

Where $\eta_b$ is the auxiliary boiler efficiency assumed to be 84% [35].

The MCR for the ship's power systems is given by the following general equation:

$$MCR_i = \frac{\dot{W}_{o,c}}{\dot{W}_{MCR,c}} \tag{10}$$

Where $\dot{W}_{o,i}$ represents the power output of each power-deliver system (*c*) (e.g. shaft power from electrical motors) provided by the SPM. The maximum shaft power output is represented by $\dot{W}_{MCR}$ [9], [36].

2.3 (b) Carbon dioxide emissions

The LNG carrier $CO_2$ emissions are calculated as follows:

$$\varepsilon_{CO2} = \sum_{i,c} M_{F\,i,c} \cdot CF_i \tag{11}$$

Where $\varepsilon_{CO2}$ is the $CO_2$ emissions produced by DF engines, boiler and GCU, *CF* is the fuel's carbon factor as given in Table 4 and *i* represents the fuel being consumed, *c* represents the fuel consumers.

2.3 (c) Overall efficiency

The LNGC overall efficiency ($\eta_T$) is described by the following equation:

$$\eta_T = \frac{\dot{W}_{out}}{\dot{W}_{in}} \cdot 100 \tag{12}$$

| Type of Fuel | RO | DO | BOG |
|---|---|---|---|
| $C_F$ ($t_{CO2}\, t_{fuel}^{-1}$)[a] | 3.206 | 3.206 | 2.750 |

[a][37]

Table 4: Carbon factors assumed per fuel type

## 2.4 OPERATIONAL MODES

Table 5 defines the different operational modes used in this work based on the Third IMO GHG Study logic [38]. In this document, the term "while at sea" groups together normal cruising, slow steaming and manoeuvring.

| Event | Variables | Logic Expression |
|---|---|---|
| Normal cruising | $\dot{W}_s$ | $\dot{W}_s$ > 65% MCR |
| Slow steaming | $\dot{W}_s$ | 20 < $\dot{W}_s$ < 65% MCR |
| Manoeuvring | $\dot{W}_s$, $v$ | $v$ > 3 kn and $\dot{W}_s$ <= 20% MCR |
| Anchored | v | 1 kn < $v$ < 3 kn |
| At berth | Shaft RPM, $v$ | $v$ < 1 kn and RPM < 1 |

* $v$ refers to ship's speed over ground

Table 5: Logical expressions defining operating modes

## 2.5 SECONDARY FILTERING

Finally, the second filtering process looked for thermal efficiencies beyond 80% and transport work (i.e. cargo transported by distance navigated) below 0. The filter did not return any observation complying with these conditions.

# 3. OPERATIONAL PROFILE

## 3.1 FLEET SAMPLE DEMOGRAPHICS

The clean dataset is made of 11 vessels with a capacity ranging between 151,000 and 173,000 m$^3$, with a median value of 158,683 m$^3$. The design draught was observed between 12.1 and 13.0 m with a median design draught of 12.5 m.

The year-of-build ranged between 2009 and 2018 without any trend indicating a growth in size through the years for this LNGC propulsive type. This may be more due to a small sample of the DFDE LNGC rather than a different behaviour to the general shipping population which has seen an increase in capacity over the past 30 years. Design speeds are 19 or 20 kn echoing the fact that, while the demand of LNG has increased (International Gas Union, 2019), the design speed is not as sensitive to time. Installed power varies between 23.0 MW and 45.0 MW with the median at 24.9 MW. There is clear correlation between installed power and design speed, design draught or capacity pointing to a preference to overpower the LNGC thanks to the modularity that this propulsion system allows.

## 3.2 TIME DISTRIBUTION AND SPEED PROFILES

On average, it is seen that the DFDE LNGC navigated 49.0% of the time in laden conditions at an averaged operational speed of 12.2 kn when accounting for time at port or 15.1 kn when just considering while at sea operations. Ballast represented 51.0% of the operational time with an average speed of 10.5 kn when accounting for time and port and 14.5 kn for operations while at sea.

| Year of build | Design speed (kn) | Design draught (m) | Capacity (m$^3$) | Propulsion power (kW) |
|---|---|---|---|---|
| 2014 | 20 | 12.1 | 151,700 | 24,950 |
| 2009 | 20 | 12.5 | 159,300 | 24,100 |
| 2009 | 20 | 12.5 | 159,300 | 24,100 |
| 2015 | 19 | 12.5 | 158,600 | 34,200 |
| 2015 | 19 | 12.5 | 158,600 | 34,200 |
| 2010 | 20 | 13.0 | 173,900 | 28,450 |
| 2013 | 20 | 12.5 | 152,900 | 23,000 |
| 2013 | 20 | 12.5 | 152,900 | 23,000 |
| 2018 | 20 | 13.0 | 169,400 | 45,000 |
| 2014 | 20 | 12.2 | 157,300 | 24,890 |
| 2014 | 20 | 12.2 | 157,300 | 24,890 |

Table 6: LNGC general description variables

Figure 2 shows the time distribution for each of the operational modes with ballast condition spending more time in activities near port such as berthing and maneuvering (i.e. 12.9%) than at laden conditions (i.e. 8.6%). One possible explanation is that at ballast legs, the temperature in the cargo tanks must be the correct one to receive LNG from the reliquefication terminals at the port or platform in order to minimise BOG generation [39], [40]. The time spent while at sea for ballast represented about 37.3% of the total time while at sea was about 40.4%. For slow steaming mode the averaged speed at ballast conditions was 13.9 kn while at laden conditions was 14.1 kn. Looking at normal cruising, the average speed at ballast conditions was 17.8 kn and for laden 17.9 kn.

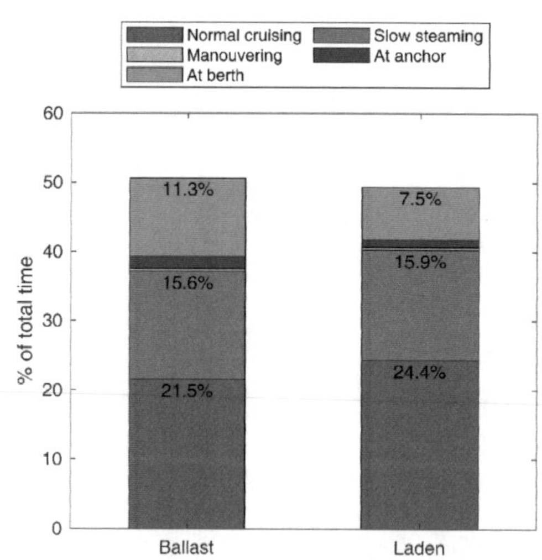


Figure 2: Percentage of time spent at each operational mode for both ballast and laden conditions

Figure 3 shows the averaged operational speed profile for the DFDE LNGC fleet observed. It is seen from this figure that at ballast there is an operational speed preference between 15.0 and 17.5 kn accounting about 24.0% of the time. For the laden condition, the preference is seen for higher speeds and sustaining those speeds for longer periods of time compared to ballast. The preferred operational speed range for laden conditions was seen between 15.5 and 20.0 kn representing about 44.3% of the time. Just about 4.0% of the time is spent at speeds above 20.0 kn.

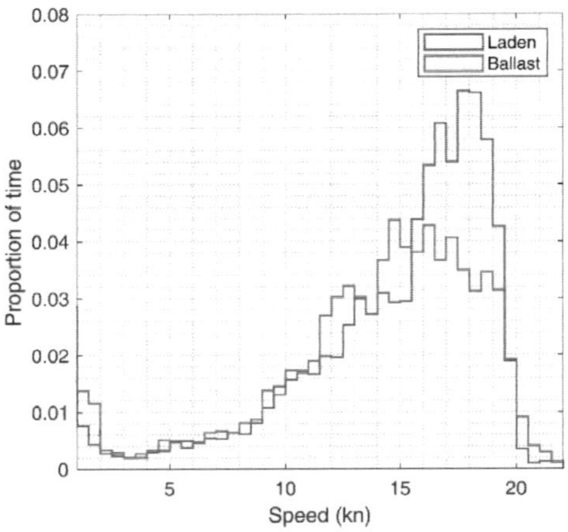


Figure 3: Average speed distribution of the total time observed for the different vessels analysed

## 3.3 FUEL, CARBON DIOXIDE EMISSIONS AND AVAILABLE POWER

Figure 4 shows total supplied power increases with ship speed as expected. At low speeds (i.e. below 6 kn), the amount of BOG disposed by the GCU dictates to maintain the tank conditions dictates the general power trend on board (see Figure 5). As the speed increases, the shaft power takes a predominant role in the power demand while less BOG is being consumed at the GCU.

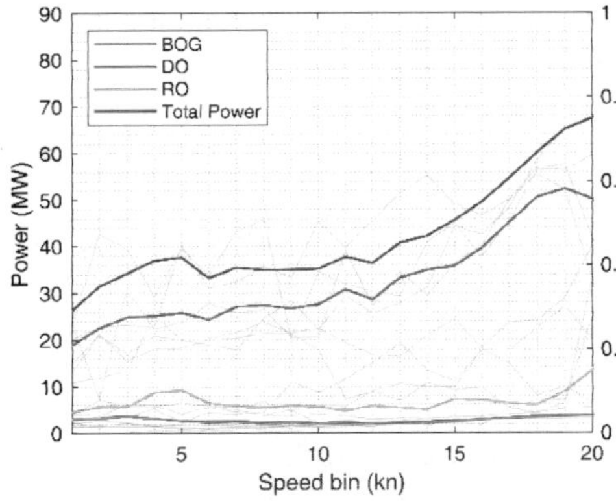


Figure 4: Average power input represented by the fuel mix per operational speed. Shaded lines represent the fuel mix power input at different operational speed for each ship

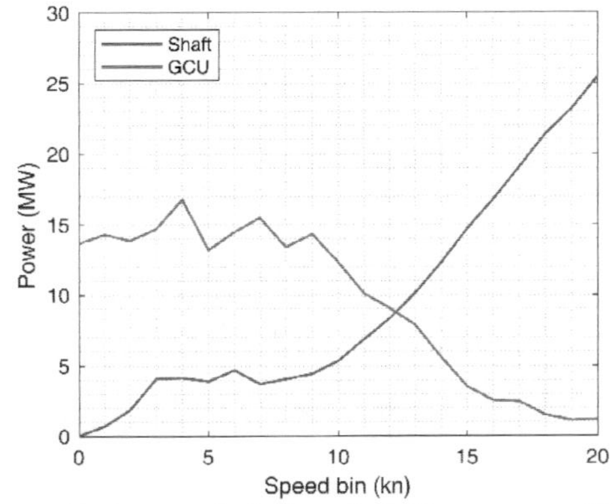


Figure 5: Averaged shaft and GCU power at different speed bins

The predominant fuel by power delivered is BOG, followed by RO and DO. At speeds above 18.0 kn it is seen that the BOG consumption starts to reduce while more prominent during ballast legs where consuming a larger proportion of RO guarantees that there is enough LNG volume that helps to maintain adequate temperatures inside the cargo tanks. In laden condition, BOG has a larger share of the fuel mix at higher speeds which helps maintain the optimum cargo conditions (i.e. pressure and temperature). The power delivery through DO consumption is relatively constant at any speed with values between 2.1 and 4.2 MW. The DO is consumed mainly as pilot fuel to help with the ignition process when the diesel engines are operating in gas mode.

Ten of the 11 ships observed had four DF engines installed on board and with the other having six. Figure 6 shows that on average DFDE LNGC operate with three engines (for 33.7% of time) while operating at speeds between 14.0 and 17.0 kn. Four and five engines are normally used for speeds above 17.0 kn and represent 32.0% of the total observed time. Single engine operation is predominantly seen for activities at berth and speeds below 2.0 kn.

Translating fuel consumption to $CO_2$ emissions gives a total of 1.20 $Mt_{CO2}$ observed which represents similar $CO_2$ emissions as Liberia or Macau in the year 2017 [41]. Similar to the total power generated on board, Figure 7 shows that at low speeds $CO_2$ emissions depend mainly on the emissions produced by the GCU and as the speed increases the shaft power demand rises the emissions up to 16.3 $t_{CO2}\,h^{-1}$.

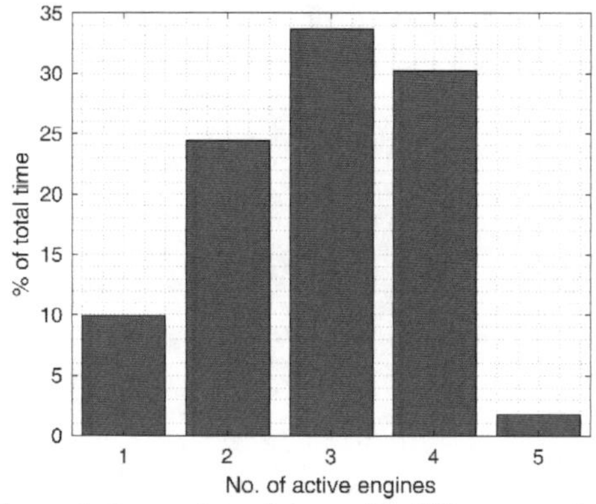


Figure 6: Proportion of time with different number of DF engines operating.

Above 18.0 kn the fuel mix starts to have a larger presence of RO instead of BOG, which gives a higher fuel mix carbon content that sustain the upward emission trend.

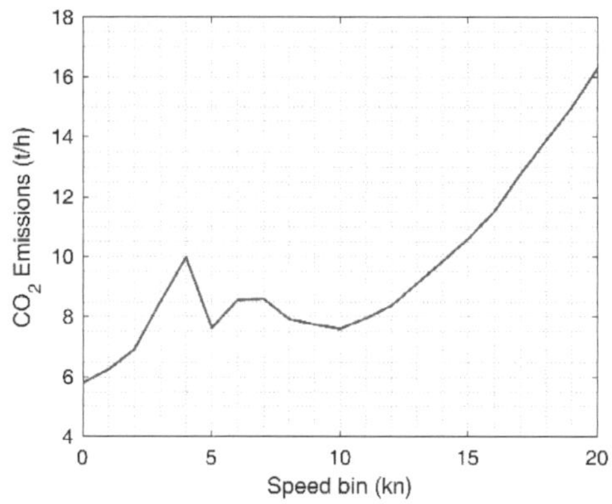


Figure 7: Average CO2 emissions per hour per operational speed bin.

The hotel load on board fluctuated between 2.4 and 3.1 MW (Figure 8). The highest power demand is seen for speed below 3.0 kn due to port related activities and at speeds above 15.0 kn caused mainly by a rise on the propulsion auxiliary machinery to support the larger power demand to the DF engines. The auxiliary boiler has its maximum power output when the ship is at berth, delivering on average 2.3 MW to operate the cargo transfer equipment. As the speed increases, the demand for steam and heating services reduces to 70 kW when the ship is navigating at 20.0 kn.

3.4 EFFICIENCY

Figure 9 shows that in general as ship speed increases, so does its overall efficiency with its highest value of 41.1% at 20.0 kn. This mainly has to do with the fact that navigating at speeds between 19.0 and 20.0 kn the DFDE LNGC reaches its design point and optimum specific fuel consumption. This implies the GCU is under minimum load since almost all the BOG is being consumed for propulsive purposes. In fact, the lowest efficiency, 25.8%, is seen at a speed of 4.0 kn which is the speeds where the GCU burns the most BOG (i.e. 16.8 MW).

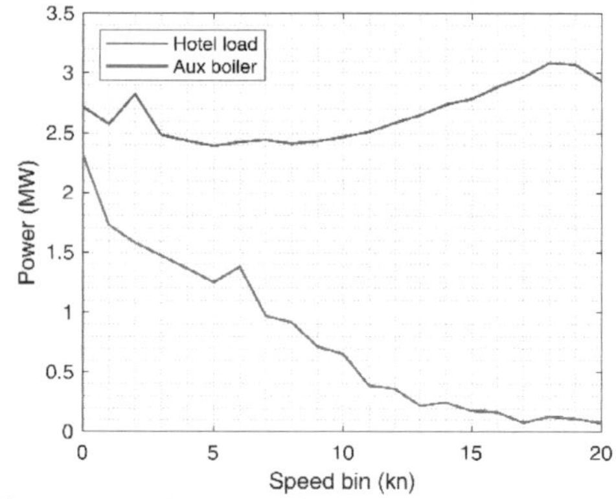


Figure 8: Hotel and auxiliary boiler load at different speed bins.

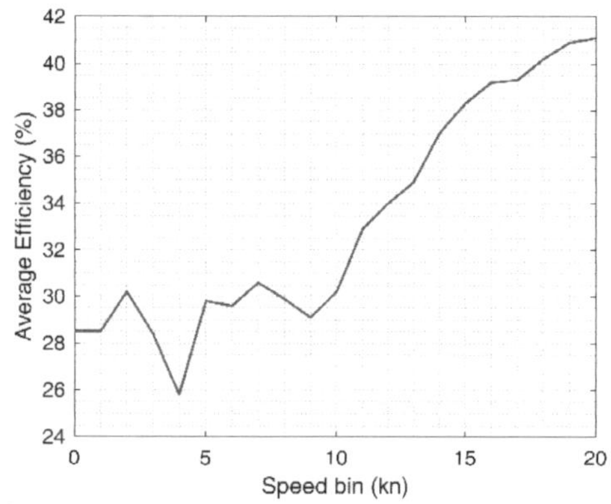


Figure 9: Overall DFDE LNGC efficiency at different operational speeds.

## 4. DATA HETEROGENEITY AND UNCERTAINTY

Due to the large dataset, results are presented in aggregated form which covers the heterogeneity of the data being recorded by the CM. While some of the fluctuations are caused by measurement uncertainty of sensors, other factors area not accounted for due to data confidentiality restrictions or non-reporting. Some of these factors are weather conditions; hull, propeller and machinery condition and maintenance routine; any hydrodynamic retrofits or changes that affect energy consumption; quality of fuel and LNG losses; cargo utilization factors; voyage chartering conditions, commercial considerations and location data that would account for geographical operating parameters.

All of these could contribute to part of heterogeneity and limitations of the study leading to not enough information to determine the degree of the effect on the outcome observed.

## 5. CONCLUSIONS

This paper aimed at exploring the operational behavior of 11 DFDE LNGC through the analysis of their CM data. This type of ships was observed operating almost similar proportion of time in their ballast and laden conditions. However, the average operational speed while at sea in ballast conditions was seen to be around 14.5 kn with preference between 15.0 and 17.5 kn. For laden conditions, the preferred speeds fluctuated between 15.0 and 17.5 kn with an average speed while at sea of around 15.1 kn.

This works provides more clarity into the power use on board for this type of ship and propulsion system. As expected, the majority of the power demand at speeds above 9.0 kn is dominated by the shaft power while at lower of speeds the GCU dictates the power consumption. The GCU has an important impact in the overall ship efficiency, and is seen to reduce efficiency to 25.8% the ship efficiency when the GCU is consuming its maximum BOG. On the other hand, at high speeds the DFDE LNGC achieves its highest efficiency of 41.1% at 20.0 kn when GCU power consumption is at is minimum.

The study quantified the $CO_2$ emissions of the 11 LNGC – 7% of the DFDE LNGC active population – for a period between the end of 2016 and beginning of 2018 which amounted to 1.20 $Mt_{CO2}$. This is an interesting reflection on how this small fleet of relatively efficient vessels can contribute considerable to GHG emissions.

Finally, this work provides more evidence for modelling fuel consumption and emissions which can help in decarbonization and energy efficient studies to decarbonize the shipping sector.

## 6. REFERENCES


[1] M. J. Economides and D. A. Wood, "The state of natural gas," *J. Nat. Gas Sci. Eng.*, vol. 1, no. 1–2, pp. 1–13, 2009.

[2] R. F. Aguilera, "The role of natural gas in a low carbon Asia Pacific," *Appl. Energy*, vol. 113, pp. 1795–1800, 2014.

[3] International Energy Agency, "Security," *IEA - Glob. Gas Secur. 2018*, p. 102, 2018.

[4] UNCTAD, "Review of Maritime Transport 2018," New York, 2018.

[5] International Gas Union, "2019 World LNG Report," Barcelona, 2019.

[6] D. Kim, C. Hwang, T. Gundersen, and Y. Lim, "Process design and economic optimization of boil-off-gas re-liquefaction systems for LNG carriers," *Energy*, vol. 173, pp. 1119–1129, 2019.

[7] E. Ekanem Attah and R. Bucknall, "An analysis of the energy efficiency of LNG ships powering options using the EEDI," *Ocean Eng.*, vol. 110, pp. 62–74, Dec. 2015.

[8] I. A. Fernández, M. R. Gómez, J. R. Gómez, and Á. B. Insua, "Review of propulsion systems on LNG carriers." 2017.

[9] Information Handling Services Markit, "IHS Maritime World Register of Ships," 2018. .

[10] IHS Markit, "IHS Markit," 2018. .

[11] H. Tu, F. Hongjun, W. Lei, and Z. Guoqiang, "Options and Evaluations on Propulsion Systems of LNG Carriers," in *Propulsion Systems*, 1st ed., IntechOpen, 2019, pp. 1–20.

[12] Wärtsilä, "Wärtsilä 50Df Engine Technology," 2009.

[13] S. M. Mousavi, R. K. Saray, K. Poorghasemi, and A. Maghbouli, "A numerical investigation on combustion and emission characteristics of a dual fuel engine at part load condition," *Fuel*, vol. 166, no. x, pp. 309–319, 2016.

[14] M. C. Cameretti, R. Tuccillo, L. De Simio, S. Iannaccone, and U. Ciaravola, "A numerical and experimental study of dual fuel diesel engine for different injection timings," *Appl. Therm. Eng.*, vol. 101, pp. 630–638, 2016.

[15] B. Wang, T. Li, L. Ge, and H. Ogawa, "Optimization of combustion chamber geometry for natural gas engines with diesel micro-pilot-induced ignition," *Energy Convers. Manag.*, vol. 122, pp. 552–563, 2016.

[16] B. Yang, C. Xi, X. Wei, K. Zeng, and M. C. Lai, "Parametric investigation of natural gas port injection and diesel pilot injection on the combustion and emissions of a turbocharged common rail dual-fuel engine at low load," *Appl. Energy*, vol. 143, pp. 130–137, 2015.

[17] M. S. Lounici, K. Loubar, L. Tarabet, M. Balistrou, D. C. Niculescu, and M. Tazerout, "Towards improvement of natural gas-diesel dual fuel mode: An experimental investigation on performance and exhaust emissions," *Energy*, vol. 64, pp. 200–211, 2014.

[18] G. Di Blasio, G. Belgiorno, and C. Beatrice, "Effects on performances, emissions and particle size distributions of a dual fuel (methane-diesel) light-duty engine varying the compression ratio," *Appl. Energy*, vol. 204, no. January, pp. 726–740, 2017.

[19] W. Li, Z. Liu, and Z. Wang, "Experimental and theoretical analysis of the combustion process at low loads of a diesel natural gas dual-fuel engine," *Energy*, vol. 94, pp. 728–741, 2016.

[20] N. N. Mustafi, R. R. Raine, and S. Verhelst, "Combustion and emissions characteristics of a dual fuel engine operated on alternative gaseous fuels," *Fuel*, vol. 109, pp. 669–678, 2013.

[21] B. J. Bora, U. K. Saha, S. Chatterjee, and V. Veer, "Effect of compression ratio on performance, combustion and emission characteristics of a dual fuel diesel engine run on raw biogas," *Energy Convers. Manag.*, vol. 87, pp. 1000–1009, 2014.

[22] R. T. Poulsen and H. Johnson, "The logic of business vs. the logic of energy management practice: Understanding the choices and effects of

energy consumption monitoring systems in shipping companies," *J. Clean. Prod.*, vol. 112, pp. 3785–3797, 2016.
[23] O. Soner, E. Akyuz, and M. Celik, "Use of tree based methods in ship performance monitoring under operating conditions," *Ocean Eng.*, vol. 166, no. July, pp. 302–310, 2018.
[24] K. Wang, X. Yan, Y. Yuan, X. Jiang, G. Lodewijks, and R. R. Negenborn, "Study on route division for ship energy efficiency optimization based on big environment data," *2017 4th Int. Conf. Transp. Inf. Safety, ICTIS 2017 - Proc.*, pp. 111–116, 2017.
[25] J. T. Hunsucker, D. Przelomski, A. Bashkoff, and J. Dixon, "Uncertainty Analysis of Methods Used to Measure Ship Fuel Oil Consumption," Fort Lauderdale, 2018.
[26] International Organization for Standardization, "ISO 19030 Ships And Marine Technology - Measurement Of Changes In Hull And Propeller Performance - Part 1: General Principles," vol. First Edit. International Organization for Standardization, p. 21, 2016.
[27] Datum Electronics Limited, "Commercial Marine Shaft Power Meter Handbook." Datum Electronics Limited, Isle of Wight, p. 12, 2018.
[28] R. N. Krikkis, "A thermodynamic and heat transfer model for LNG ageing during ship transportation. Towards an efficient boil-off gas management," *Cryogenics (Guildf).*, vol. 92, no. April, pp. 76–83, 2018.
[29] International Maritime Organization, "RESOLUTION MEPC.245(66) - 2014 GUIDELINES ON THE METHOD OF CALCULATION OF THE ATTAINED ENERGY EFFICIENCY DESIGN INDEX (EEDI) FOR NEW SHIPS." International Maritime Organization, London, p. 30, 2014.
[30] G. G. Dimopoulos and C. A. Frangopoulos, "A Dynamic Model for Liquefied Natural Gas Evaporation During Marine Transportation," *Int. J. Thermodyn.*, vol. 11, no. 3, pp. 123–131, 2008.
[31] International Maritime Organization, "Resolution MEPC.281(70): Amendments to The 2014 Guidelines on The Method of Calculation of The Attained Energy Efficiency Design Index (EEDI) for New Ships (Resolution MEPC.245(66), As Amended By Resolution Mepc.263(68))," *Annex 9*. International Maritime Organization, London, pp. 1–10, 2016.
[32] M. Lam, "Marine Fuel Oil ( MGO & HFO ) and FCC Additive Technical Analysis," 2017.
[33] International Gas Union, "Natural Gas Conversion Guide." International Gas Union, Oslo, pp. 19–21, 2012.
[34] F. Li, "Discussion on Improving Marine Boiler Efficiency on Board," *Adv. Intell. Syst. Res.*, vol. 156, no. Meici, pp. 907–910, 2017.
[35] A. . Karthikeyan, "Energy Performance Assessment of Boilers," 2014.
[36] Kyma AS, "Internal Communications." Kyma AS, p. 10, 2018.
[37] International Maritime Organization, "Guidelines For Voluntary Use Of The Ship Energy Efficiency Operational Indicator (EEOI)," London, 2009.
[38] International Maritime Organization, "Third IMO Greenhouse Gas Study 2014," London, 2014.
[39] Y. Shao, Y.-H. Lee, Y.-T. Kim, and H.-K. Kang, "Parametric Investigation of BOG Generation for Ship-to-Ship LNG Bunkering," *J. Korean Soc. Mar. Environ. Saf.*, vol. 24, no. 3, pp. 352–359, Jun. 2018.
[40] J. Lu, S. Xu, J. Deng, W. Wu, H. Wu, and Z. Yang, "Numerical prediction of temperature field for cargo containment system (CCS) of LNG carriers during pre-cooling operations," *J. Nat. Gas Sci. Eng.*, vol. 29, pp. 382–391, Feb. 2016.
[41] M. Crippa *et al.*, "Fossil CO2 emissions of all world countries : 2018 report.," European Commission, Brussels, 2018.

## 7. AUTHORS BIOGRAPHY

**Jean-Marc Bonello** is a Post-Doctoral Researcher at the UCL Energy Institute and a consultant at UMAS International Ltd. with a research interest in information and data use in decision-making related to the uptake of energy efficiency technologies in shipping towards emission reduction. Through UCL, he has worked with various owners and operators to understand the role that AIS, in-service and metadata analytics has in their daily operations and its application for appraising future decisions and investments. He is a lead author of the Sea Cargo Charter and an author of the Fourth IMO Greenhouse Gas Study. He has an MSc in Marine Engineering and Naval Architecture from the University of Southampton and a BEng in Mechanical Engineering from the University of Malta. His work experience includes design and consultancy in mechanical building services as well as naval architecture EPC consultancy in the oil and gas industry.

**Carlos González Gutiérrez** (MSc Marine Engineering and BSc Naval Architecture) specializes in propulsion systems. He is a PhD candidate within marine engineering and shipbuilding department of University of Cantabria, a 3rd Engineer according STCW-95 and currently working as Technical production manager at Kyma As.

**Dr Santiago Suárez de la Fuente** is Lecturer in Energy and Transport at the UCL Energy Institute. His research focuses on marine energy efficient technologies in particular waste energy recovery systems (WERS), holistic modelling approaches for the maritime sector, performance data analytics and decarbonisation pathways. He is co-developer of the ship design software called Whole Ship Model (WSM), which has been used successfully to inform different maritime groups of the implications of different decarbonisation approaches. . Dr Suarez de la Fuente did his PhD at UCL on marine waste heat recovery systems (WHRS). He was granted the Institute of Marine Engineering, Science and Technology (IMarEST) Stanley Gray fellowship for his research, and was awarded the IMarEST's prestigious Denny Medal in 2017. He studied his bachelor's degree in Mechanical Engineering at the Institute of Technology and Higher Education (ITESO, Mexico).

# AN INVESTIGATION ABOUT ADDED POWER IN HEAVY WEATHER FOR PASSENGER SHIPS WITH FOCUS ON WIND EFFECT

**A Boveri, M Cassola** and **P Gualeni** University of Genoa, Italy


## SUMMARY

The passenger ships propulsion performance in heavy weather condition has been investigated, using full scale data recorded on board. An "added power" coefficient has been identified, inclusive of both wind and waves effects, in relation also with the frequency of occurrence for each relative wind direction value. Moreover, due to the importance of the topside exposed surface, typical feature of passengers ships (e.g. cruise ships and ferries), attention has specifically been paid to the wind contribution. An attempt to derive the solely effect of wind action from a full scale data-sample has been finalized; in this perspective non-dimensional coefficient of wind force has been derived and plotted as a function of wind relative direction angle.


## 1. INTRODUCTION

Energy efficiency performance plays an increasingly significant role during the ship design process for its implications in terms of both costs and International Maritime Organization (IMO) rules compliance. The increased awareness about potential air pollution from ship exhaust gases has in fact driven the development of stringent international rules, which in turn establish an effective flywheel for knowledge enhancement, advances in innovative solutions and further marine technology developments.

The aim of this work is to gain a further insight into the influence of heavy weather conditions on ships in service, with a specific focus on wind effect. A better understanding of this aspect can in fact support the appropriate selection and sizing of the power generation and propulsion system on board. Traditionally an increment beside the ship resistance in calm water is to be taken into account, due to environment phenomena like waves and wind [1, 2].

It is state of the ar" to think in terms of additional power and the topic is framed with the nomenclature of sea margin. The increment of power is commonly calculated between the 10 and 25 percent of the propulsion power evaluated with reference to the calm water condition [3].

Nevertheless, a better accuracy in the prediction of the so called added power is more and more necessary, also in relation with the IMO rules compliance. Investigations about ship performance in operations, i.e. sailing in waves and winds, can help to this aim and a very important asset in this perspective is embedded in the huge amount of data that nowadays are recorded on board. The adequate understanding of such source of information can be a precious feedback for the design of new ships [4, 5] i.e. can be transformed in knowledge. As a further step, it can also represent the domain where develop route optimization policies, useful for an efficient ship operational life [6].

With a special attention to wind effects, it is recognized that container ships have significant exposed areas above the waterline, especially when fully loaded of containers and several studies are available on relevant wind forces, based both on numerical and experimental approaches [7, 8, 9, 10]. CFD studies are increasingly present in literature on the topic of wind effect [11] and comparisons of results with wind tunnel data and in situ full scale wind speed measurements have shown that CFD is a valid research tool to investigate the mean air flow over ships [12]. Beside containerships, several other ship typologies are investigated [13, 14] but very few data are available in relation with wind effect on passenger ships [15].

As already mentioned, in this paper the ship aerodynamic performance is investigated elaborating from data which are the result of wind action, with a specific focus on passenger ships because of the large exposed areas typically characterizing Ferries and Cruise ships.

At the same time, a better understanding is also advantageous from a safety point of view, especially when wind comes from beam sides.

Finally, when speaking about wind effect on ships and energy efficiency, it is worth mentioning that wind can also represent a possible source of energy, as evidenced in the literature about the wind effect for "wind assisted" ship propulsion, an emergent topic in the context sustainable transport [16, 17]. Therefore in general the topic is likely to become more and more appealing and under the attention of many ship owners.

## 2. AN OVERVIEW ABOUT EVALUATION OF WIND FORCES ON SHIPS

In literature it is possible to find several methodologies enabling the study of the aerodynamic performance of ships, on a 360° direction perspective.

An obvious general premise is that, due to the ship's forward speed, the wind influence is to be considered in terms of relative wind intensity and direction.

The longitudinal component of the wind force can be opposite the ship's heading, therefore being detrimental for ship efficiency. At the same time the transverse wind force causes yaw, drift and deviation from the intended course, which can cause, beside possible stability issues, added resistance. However, the longitudinal force remains the most important.

The relative wind field experienced by the ship topside is also influenced by a combination of the wind field with

 

its natural vertical wind gradient and the homogenous wind field [18,19, 20].
The effect of wind has been investigated by several authors along the decades and a brief overview is reported here.
Isherwood [21] proposed parametric expressions for wind forces and moments coefficients, derived from regression analysis based in turn on previous experimental data: 36 coefficients are given in terms of equations depending on the main characteristics of the ship topsides. This method can be applied to various merchant ships types (mainly tankers and bulk carriers).
A possible shortcoming is that the experimental records are dated back in 1973 and therefore it is likely they refer to obsolete topside geometries, not suitable for new generation ship profiles and even less for passenger ships.
Some years later Van Berlekom [22] proposed a method based on wind tunnel tests data, performed on tankers, containers and Ro/Ro ships models. The squared ship length was used as representative of the exposed reference area and the wind coefficient was formulated as function of the ratio between the actual projected area and the squared ship length. Characteristic curves of wind coefficient in relation with wind direction are available. Van Berlekom's method results are usually about 10% higher than the Isherwood method's ones. At the same time the application of the method is not suitable for passenger ships, because it is based on cargo ship geometries.
More recently Blendermann [23] presented a systematic data collection about wind forces. The outcome is the availability of formulas for the calculation of wind non dimensional coefficients for longitudinal and transverse forces and for yaw and roll moments. Such formulas are functions of: drag coefficient, angle of attack, frontal and transverse areas exposed to wind, coordinates of the centre of area and a parameter called the "cross-force parameter". This parameter is obtained statistically from experimental data.
The methodology proposed by Townsin et al. [24] addresses the problem in a different way in comparison with the above mentioned ones. In fact, in this case a correction factor related to the presence of waves is also considered and therefore a more comprehensive attention to the weather condition effect on the ship resistance is given. The work was developed on the study of some Series 60's ship models in laden and ballast conditions, at five Froude numbers from 0.10 to 0.30, over the range of four ship lengths (100m to 400m) and for a range of weather conditions represented by Beaufort Numbers $\leq$ 7.
At present, when in the need to evaluate the wind effect on ship performances during sea trials, the ITTC [25] recommends the method of Fujiwara et al. [26] to calculate forces and moments due to wind acting on a ship.
In Fujiwara et al., [27] a methodology is developed for the estimation of wind forces and moments acting on large passenger ships, based on a physical-component-model of the wind loads. In this paper a particular relevance will be given to the longitudinal force coefficients $C_{AX}$.

## 3. ANALYSIS OF RECORDED DATA

This paper analyses the propulsion system performances of two passenger ships, i.e. a Cruise ship and a Ferry, with reference to the weather parameters, derived from field data monitored and recorded on board during navigation.
In the following, a statistical analysis of different aspects defining the ships operational profile and weather conditions will be described in order to put in evidence the most representative parameters.
It might be not straightforward to obtain satisfactory description of environmental data (wind and waves) from the onboard data record systems. On the other hand, technology is in continuous evolution and the increasing introduction of the wave's radar system on board surely represents a significant improvement. In fact one of the biggest challenge in this kind of investigations is the reliable measurement of wave characteristics.
A possible practical approach, waiting for the proper technology diffusion, could be to rely on the relation between sea condition and the wind state as traditionally meant. This approach is very common and for example already applied in principle when dealing with the JONSWAP and Pierson-Moskowitz's (PM) wave spectra analytical formulation. In these cases, in fact, it is possible to represent the significant wave height as a function of the wind intensity [28, 29]. The PM spectrum was originally proposed for fully-developed sea. The JONSWAP spectrum extends PM to include fetch limited seas, describing sea states under development. Both spectra make reference to wind sea conditions that often occur for the most severe sea-states, that are the focus of interest in this work.

### 3.1 ANALYSIS ON SEASTATE PARAMETERS

However, as a basic approach, in this paper the Beaufort scale has been considered for analysing and organizing the available full-scale data. The Beaufort scale is effective in describing the comprehensive weather conditions, since it is graduated as a function of wind intensity, but it also relates to a sea-state.
A statistical description of the environmental conditions in terms of Beaufort Scale has been carried out for the two ships.
Actually, from the available full scale data, several very interesting information about the real operational conditions of the ships can be derived. It is possible for example to study the probability density function (pdf) distribution of the wind absolute speed (or "true wind speed") and an example is given in Fig. 1.

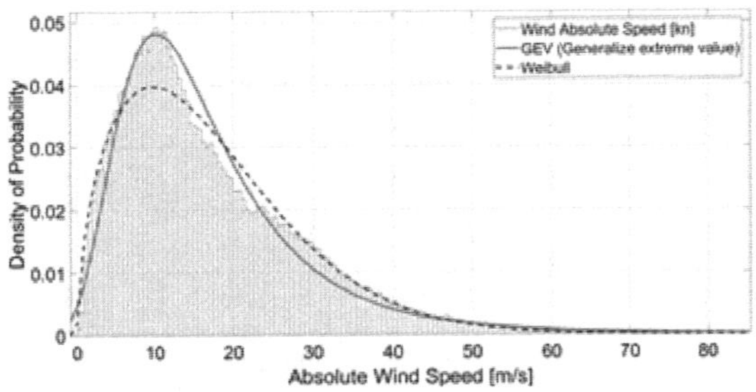


Figure 1: Absolute Wind Speed Intensity pdfs from the Data-Samples for the Ferry

A GEV (Generalize Extreme Value) distribution appears as more suitable to describe the mean value but the Weibull distribution is better to describe the higher absolute wind speed values that are actually the focus of this study [30].

Another feature detected by the on-board measuring system is the absolute wind direction or "true wind direction".

Absolute or true weather data provided by ship records are very interesting for meteorological observations and investigation; in this perspective ships can be considered a very convenient weather station.

However for the ship energy efficiency, the relative wind parameters (relative wind speed and relative wind direction) are the relevant ones, since they describe the wind and waves as perceived by the ship.

The wind relative parameters are dependent on other quantities represented in Fig. 2.

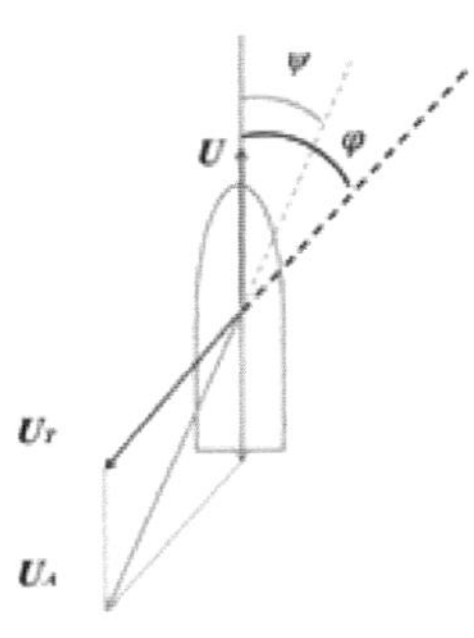


Figure 2: Relative Wind Components

Where:

U is the ship speed;
$U_T$ is the absolute wind speed;
$U_A$ is the relative wind speed;
Ψ is the relative wind direction in the ship's reference system (Oxyz);
φ is the absolute wind direction in the ship's reference system (Oxyz).

The pdfs of wind relative speed for the Ferry are shown as an example in Fig. 3.

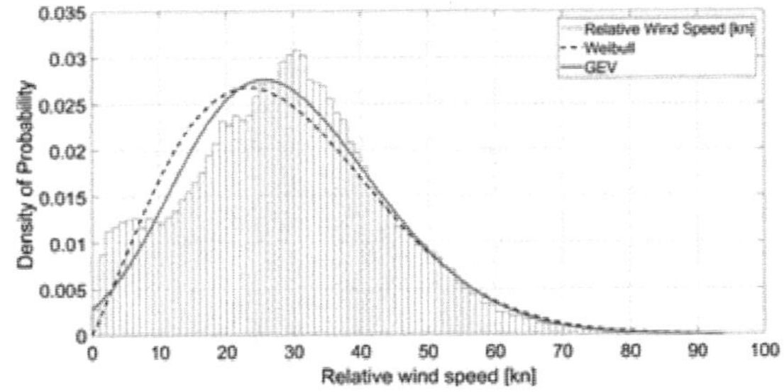


Figure 3: Relative Wind Speed Intensity pdfs from the Data-Samples for the Ferry

Also in this case, the two Analytical pdf formulations (i.e. GEV and Weibull) are represented.

## 3.2 ANALYSIS ON SHIP'S PARAMETERS

The ship's operational parameters considered for the investigation are: speed, displacement and propulsion power.

Their statistical analysis is important in order to evaluate their modal values, selected as reference parameters able to be representative for a typical operational condition.

The analysis of the ships speeds shows two modal values for the Ferry (around 22-23 kn and 26-27 kn) as shown in Fig. 4; even three are detected for the Cruise (around 13-14 kn, 18-19 kn and 21-22 kn) as shown in Fig. 5.

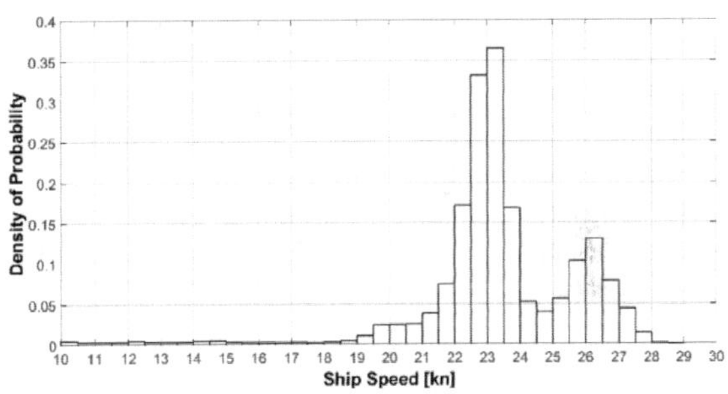


Figure 4: Ship Speed [kn] pdf for the Ferry

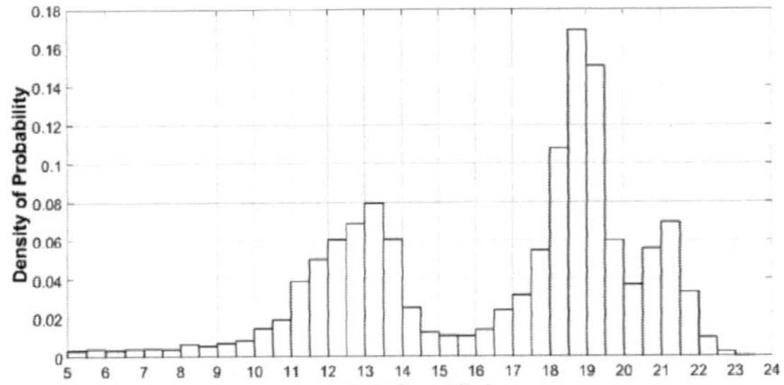


Figure 5: Ship Speed [kn] pdf for the Cruise Ship

For the selection of a representative displacement, the two ship typologies need different approaches: the Ferry in fact is characterized by a rather large range of values during the operational activity and for the purpose of this investigation a reference value has been assumed as 23800 t; on the other hand the Cruise ship operates nearly at a constant displacement (≈ 56500 t) so in such case it was not necessary a statistical analysis. The distribution of the respective propulsive power, after filtering the data-sample for the reference displacement, is shown in Fig. 6 and Fig. 7.

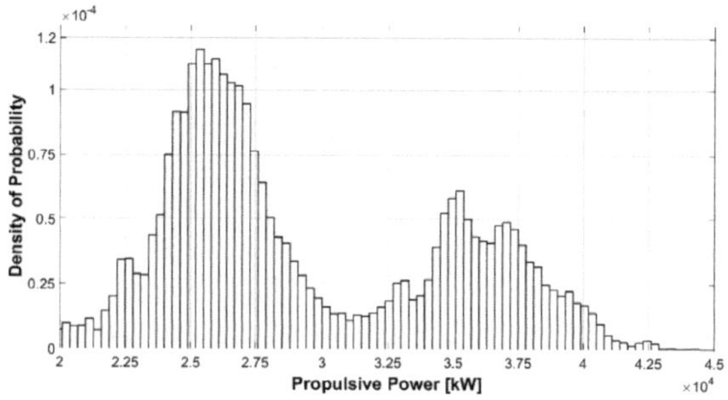


Figure 6: Propulsive Power pdf for the Ferry

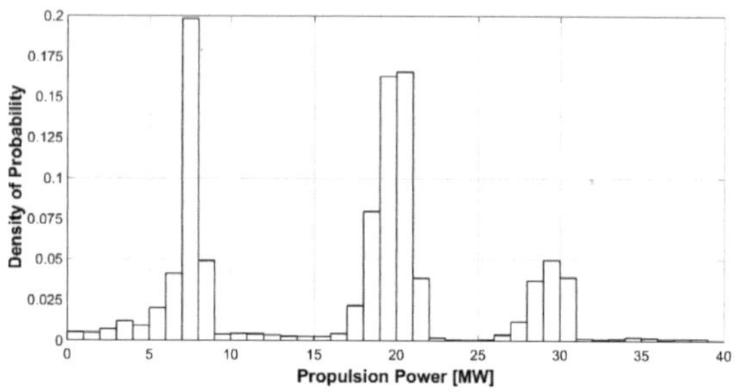


Figure 7: Propulsive Power pdf for the Cruise Ship

The modal values relevant to propulsion power for the Ferry can be assumed 25.5 and 36 MW; instead for the Cruise ship the selected values are 8, 20 and 30 MW.

## 4. ADDED POWER DUE TO HEAVY WEATHER FOR WIND AND WAVES

The two vessels performance (in terms of propulsion power) is then evaluated under the effect of adverse weather conditions.

As a first step, a baseline, i.e. a reference condition, is identified as the situation where the weather does not affect the performance. In this perspective, the data sample has been divided according to BN i.e. according to wind speed at first and then in turn in terms of the associated the sea-state, as provided in the BN tables.

It has been observed that, there is no significant influence until the level BN5. Nevertheless, when the BN exceeds level 7, this effect is found to be quite significant.

The effect of the relative wind direction, above BN5 level, has been put in evidence as well and to this aim the data-sample is divided in relation with the different wind relative angles (ψ).

Ships in scheduled services tend to operate at constant speed and need a sufficient power margin to reasonably guarantee the necessary speed at any weather condition. We will call "added power" the additional ship propulsion power required in heavy weather conditions. In the present paper, the influence of wind and waves on ship speed can be estimated by filtering the propulsion power data-sample for different weather conditions (BNs) and relative wind direction (ψ).

To evaluate the added power value, the data-sample relevant to BN from 5 to 9 level are considered and are divided in terms of relative wind direction in a more refined way as already mentioned.

For Ferry

- 0-10°, 10-20°, 20-30°, 40-50°, 50-60°, bow directions
- 60-80°, 80-90°, 90-110°,110-130°, beam directions
- 130-150°, 150-165° e 165-180°, stern directions.

For Cruise

- 0-10°, 10°-20°, 20°-30°, 30°-40°, 40°-60°, bow directions
- 60°-80°, 80°-90°, 90°-110°, 110°-130°, beam directions
- 130°-180°, for stern directions.

It was no possible to take the same discretization in terms of ψ for the two ships because of the differences in the data-sample, collected with different modality.

The added power due to adverse weather is then identified as a difference between values of required power in heavy weather condition (BN from 5 to 9) and the value found for the calm weather state.

The typical trend of the added power due to heavy weather in relation with the relative wind direction is shown in Fig. 8.

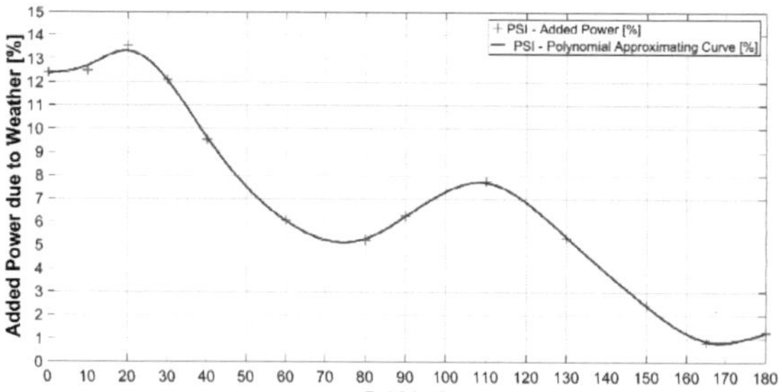


Figure 8: Added Power (%) Trend for the Ferry

### 4.1 AN OVERALL SEA MARGIN

In principle, in order to correctly asses in a design phase the influence of weather, the power increment or speed loss should be judged in relation with the occurrence frequency of the specific weather conditions (i.e. the occurrence of BN) [18], that in turn can be related to the ship's routes.

Starting from the knowledge in advance of all the significant information i.e. for each BN: the added power as a function of the relative angle of the wind, the probability of meeting a specified relative wind direction the probability of occurrence of each weather condition (BN) and moreover the sea trial power required in calm weather, it is possible to derive what can be a more ideal and rational formulation of the Sea Margin.
In this paper the evaluation is focussed on the actual situation i.e. the one already experienced by the ship and evidenced by the recorded data. The kind of results in this case can help in understanding a possible fuel consumption increment spotted afterward.
Inspired by a method proposed by Molland et al. [19], the weighted average power increase is calculated as:

$$\sum (power\ increase \cdot frequency) = \sum (\Delta P \cdot \sigma)$$

Where, in this case the frequency **σ** is the probability of occurrence of each relative wind direction value, found from the ship recorded data.
The overall sea margin in relation with the data-sample base-line is now evaluated in Table 1 and Table 2 respectively for the Ferry and the Cruise ship.
$P_{ADDED}$ is the added power expressed as a percentage of the calm water power. It is the same information reported in Fig. 8.

Table 1 - Evaluation of the Sea Margin with Data-Sample Base-Line for the Ferry

| **Ψ [deg]** | **$P_{ADDED}$ (%)** | **σ** | **σ·$P_{ADDED}$(%)** |
|---|---|---|---|
| 0 | 12,5 | 0,057 | 0,713 |
| 0-10 | 12,6 | 0,255 | 3,213 |
| 10-20 | 13,5 | 0,2168 | 2,927 |
| 20-30 | 12 | 0,146 | 1,752 |
| 30-40 | 9,5 | 0,086 | 0,817 |
| 40-60 | 6 | 0,089 | 0,534 |
| 60-80 | 5,2 | 0,0478 | 0,249 |
| 80-90 | 6,2 | 0,0214 | 0,133 |
| 90-110 | 7,5 | 0,039 | 0,293 |
| 110-130 | 5,3 | 0,0163 | 0,086 |
| 130-150 | 2,4 | 0,0093 | 0,022 |
| 150-165 | 0,8 | 0,0109 | 0,009 |
| 165-180 | 1,3 | 0,0105 | 0,014 |
| **TOT** | | **1,0** | **10,7** |

As a result, in this case, the values the overall sea margin calculated as weighted average are:

- ≈10% for the reference speed of 23 knots for the ferry and
- ≈17% for the Cruise at a speed of 19 knots.

Table 2 - Sea Margin Evaluated in comparison with the Data-Sample Base-Line for the Cruise Ship

| **Ψ [deg]** | **$P_{ADDED}$ (%)** | **σ** | **σ·$P_{ADDED}$(%)** |
|---|---|---|---|
| 0 | 31,0 | 0,03 | 0,9296 |
| 0-10 | 31,2 | 0,14 | 4,3733 |
| 10-20 | 29,3 | 0,12 | 3,5210 |
| 20-30 | 27,5 | 0,11 | 2,8870 |
| 30-40 | 22,2 | 0,09 | 1,9985 |
| 40-60 | 11,9 | 0,14 | 1,6627 |
| 60-80 | 2,2 | 0,12 | 0,2635 |
| 80-90 | 2,6 | 0,06 | 0,1453 |
| 90-110 | 8,7 | 0,07 | 0,6425 |
| 110-130 | 8,3 | 0,05 | 0,4142 |
| 130-150 | 3,8 | 0,04 | 0,1327 |
| 150-180 | 0,4 | 0,04 | 0,0180 |
| **TOT** | | **1,0** | **17%** |

## 4.2 NON-DIMENSIONAL WIND COEFFICIENT

In the previous paragraph, the added power due to combination of waves and wind together has been evaluated, without separating the two effects. In the following, a specific attention is given instead to the evaluation of the added power component due only to wind. The high number of data-samples enables a couple of considerations.
Thanks to the application of Fujiwara et al. [27] methodology, it is possible to calculate the longitudinal force coefficients $C_{AX}$ for the Ferry and to observe, in the case of the Ferry for example, that for an angle of relative wind direction around 70°, the added power component due to wind can be considered negligible. Therefore, at this angle, the added power can be assumed as totally due to wave effect and this value can be kept constant (as an approximation) to derive, by subtraction, the wind effect value for all the other relative wind directions.
On the other side a second option could be considered to quantify the added power component due only to wave, to be used exactly in the same way: within sea-state data-sample from BN 5 to BN 9, only data with relative wind velocity close to zero (UA < 5 kn) can be selected. The relevant power records are then identified and averaged and the result is assumed as the representative added power component due to the waves.
From the two different approaches described above, we have found very similar values of the added power due to waves (i.e. around 27500 kW) Therefore we selected such figure to derive in turn the added power due to wind for the different relative directions as shown in the following Table 3 and Fig. 9. The same whole procedure has been followed for the Cruise ship.

Table 3 - Evaluation of the Added Power in due to Wind for the Ferry

| $\psi$ | $P_{BN5\text{-}9}$ | $P_{CALM\ WEATHER}$ | $P_{ADDED\ TOT}$ | $P_{AT\ 67° (BN5\text{-}9)}$ $P_{ADDED\ WAVES}$ | $P_{ADDED\ WIND}$ |
|---|---|---|---|---|---|
| deg | kW | kW | kW | kW | kW |
| 0 | 29534 | 26277 | 3257 | 27722 | 1811 |
| 10 | 29560 | 26277 | 3283 | 27722 | 1838 |
| 20 | 29836 | 26277 | 3559 | 27722 | 2113 |
| 30 | 29447 | 26277 | 3170 | 27722 | 1724 |
| 40 | 28781 | 26277 | 2504 | 27722 | 1059 |
| 60 | 27865 | 26277 | 1588 | 27722 | 142 |
| 80 | 27647 | 26277 | 1370 | 27722 | -75 |
| 90 | 27920 | 26277 | 1643 | 27722 | 197 |
| 110 | 28312 | 26277 | 2035 | 27722 | 589 |
| 130 | 27674 | 26277 | 1397 | 27722 | -49 |
| 150 | 26917 | 26277 | 640 | 27722 | -805 |
| 165 | 26497 | 26277 | 220 | 27722 | -1226 |
| 180 | 26613 | 26277 | 336 | 27722 | -1110 |

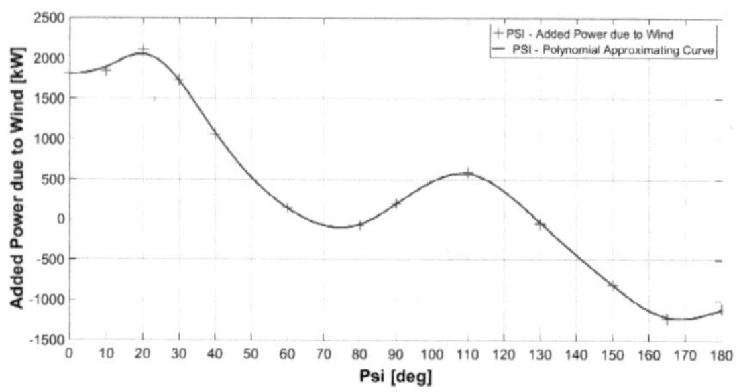


Figure 9: Added Power due to Wind for the Ferry

Then, correspondent wind added resistance has been calculated as reported in equation:

$$R_{added,\ wind} = \frac{P_{added,\ wind}}{V_{ship}}$$

where $V_{ship}$ is the reference ship speed (e.g. 23 knots for the Ferry).

Once the wind added resistance is evalutated, it is possible to obtain the non-dimensional coefficient for force due to wind.

$$C_{AX,\ wind} = \frac{R_{added,\ wind}}{1/2\rho_{air} A_X V^2}$$

Where:

$R_{added,\ wind}$ is the added resistance due to wind [kN]

$\rho_{air}$ is the density of the air [kg/m$^3$]

$A_X$ is the transversal exposed area of the ship in [m$^2$]

V is the wind speed in [m/s].

The result for the Ferry is reported in Fig. 10. For the Cruise it was not possible to derive the same curve because of the lower number of observations in following wind direction.

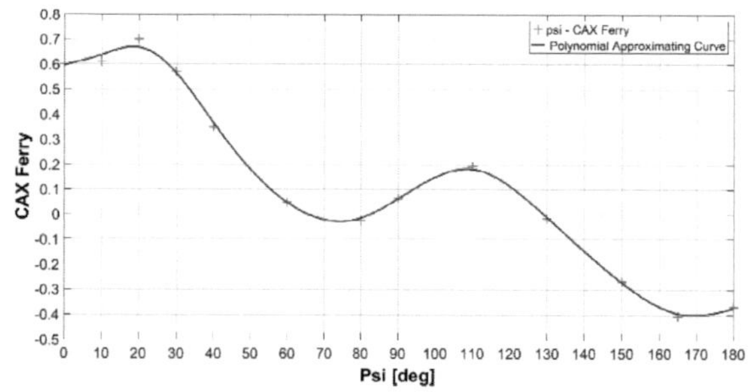


Figure 10: Longitudinal Non-dimensional Coefficient $C_{AX\_wind}$ for Ferry

This non-dimensional coefficient is comparable to the one obtained with the method of Fujiwara et al. as shown in Fig. 11. For this reason, it can be deduced that the methodology applied to the ships under exam can be reliable for the evaluation of the added resistance due to an adverse weather condition and/or also for the wind effect on ship performance.

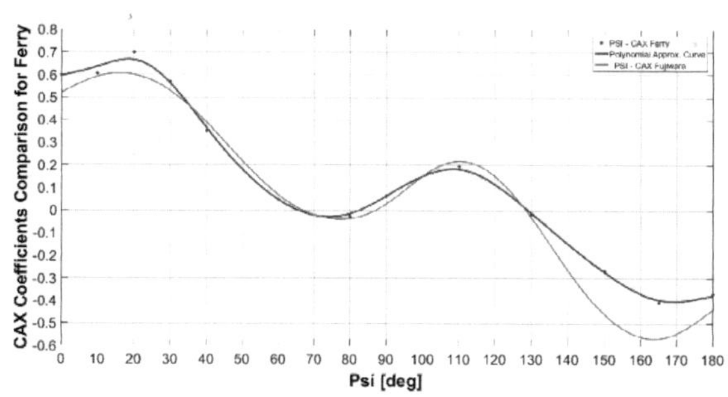


Figure 11: Comparison between the Fujiwara Longitudinal Non-dimensional Coefficient $C_{AX\_wind}$ and the One Calculated through the Data-Sample for the Ferry

## 5. CONCLUSIONS

Especially for ships with significant exposed topside areas, the wind effect can assume a relevance worthy to be investigated. At the same time, most of the studies found in literature to assess the wind added resistance deal with cargo ships like oil tankers, bulk carriers and container ships, therefore to add some data about passenger ships can be useful.

In this paper, a large sample data, derived from the on-board monitor and recording system, has been analysed to understand the performance of the vessels in different weather conditions. Further, such data-sample has been used to identify a method for added power evaluation due to adverse weather conditions, in relation with the ships routes, based on a weighed sea margin.

Finally, the dimensionless wind force coefficient $C\_{AX\_wind}$ has been derived from the same data-sample recorded on board, useful to characterize the ship performance. Such non-dimensional coefficients are usually obtained only through experimental tests in towing tanks or in the wind tunnels.

It has been shown how suitable and robust can be the formulation provided by Fujiwara for the calculation of the longitudinal force provided by wind.

However, it is worth mentioning that the data recorded on board might be comprehensive of various effects (e.g. current, fouling…) therefore the sea margin as it has been calculated is actually intended in an extended meaning.

## 6. REFERENCES


1. SHIGUNOV, V., BERTRAM, V., 'Prediction of added power in seaway by numerical simulation', *Proceedings of the 9th Int. Conf. on High-Performance Marine Vehicles HIPER 2014*, Athens, Greece, 2014.
2. SHIGUNOV, V., 'Added power in seaway', *Ship Technology Research. 64, 65-75*, 2017.
3. ABS, 'Ship energy efficiency measures advisory', 2013.
4. TSUJIMOTO, M., ORIHARA, H., 'Performance prediction of full-scale ship and analysis by means of onboard monitoring (Part 1 ship performance prediction in actual seas)', *Journal of Marine Science and Technology. 1-18*, 2018.
5. ORIHARA, H., TSUJIMOTO, M., 'Performance prediction of full-scale ship and analysis by means of on-board monitoring (Part 2: Validation of full-scale performance predictions in actual seas)', *Journal of Marine Science and Technology. 23, 782-801*, 2018.
6. LI, X., SUN, B., ZHAO, Q., LI, Y., SHEN, Z., DU, W., XU, N., 'Model of speed optimization of oil tanker with irregular winds and waves for given route', *Ocean Engineering. 164, 628-639,* 2018.
7. LUO, S., MA, N., Hirakawa, Y., 'Evaluation of Resistance Increase and Speed Loss of a Ship in Wind and Waves', *Journal of Ocean Engineering and Science. 1*, 2016.
8. ANDERSEN, I., 'Wind on Post-Panamax Containership' *Ocean Engineering. 58, 115-134*, 2013.
9. JANSSEN, W.D., BLOCKEN, B., VAN WIJHE, H.J., 'CFD simulations of wind loads on a container ship: Validation and impact of geometrical simplifications', *Journal of Wind Engineering and Industrial Aerodynamics. 166, 106-116*, 2017.
10. VAN HE, N., MIZUTANI, K., IKEDA, Y., 'Reducing air resistance acting on a ship by using interaction effects between the hull and accommodation', *Ocean Engineering. 111, 414-423,* 2016.
11. SAYDAM, A. Z., TAYLAN, M., 'Evaluation of wind loads on ships by CFD analysis', *Ocean Engineering. 158, 54-63*, 2018.
12. MOAT, B. I., YELLAND, M. J., PASCAL, R.W., MOLLAND, A. F., 'The effect of ship shape and anemometer location on wind speed measurements obtained from ships', *The Royal Institution of Naval Architects. 133-139*, 2005.
13. VALČIĆ, M., PRPIĆ-ORŠIĆ, J., 'Hybrid method for estimating wind loads on ships based on elliptic Fourier analysis and radial basis neural networks', *Ocean Engineering. 122, 227-240*, 2016.
14. FUJIWARA, T., UENO M., NIMURA, T., 'Estimation of Wind Forces and Moments acting on ships', *Journal of the Society of Naval Architects of Japan, 77-90*, 1998.
15. UENO, M., KITAMURA, M., SOGIHNARA, N., FUJIWARA, T., 'A Simple Method to Estimate Wind Loads on Ships', *Proceedings of the World Congress on Advances in Civil, Environmental, and Materials Research (ACEM' 12)* Seoul, Korea, 2012.
16. BENTIN, M., ZASTRAU, D., SCHLAAK, M., FREYE, D., ELSNER, R., KOTZUR, S., 'A New Routing Optimization Tool-influence of Wind and Waves on Fuel Consumption of Ships with and without Wind Assisted Ship Propulsion Systems', *Transportation Research Procedia, 14, 153-162*, 2016.
17. VIOLA, I. M., SACHER, M., XU, J., WANG, F., 'A numerical method for the design of ships with wind assisted propulsion', *Ocean Engineering. 105, 32-42,* 2016.
18. ARRIBAS, F., 'Some Methods to Obtain the Added Resistance of a Ship Advancing in Waves', *Ocean Engineering*. 34, 946-955, 2007.
19. MOLLAND, F. A., TURNOCK R. S., HUDSON, A. D., 'Components of Hull Resistance in Ship Resistance and Propulsion, practical estimation of ship propulsive power', *Cambridge University Press,* 2011.
20. FUJIWARA, T., UENO, M., IKEDA, Y., 'Cruising Performance of a Large Passenger Ship in Heavy Sea', *Proceeding of the 16th International Offshore and Polar Engineering Conference*, San Francisco, USA, 2006b.
21. ISHERWOOD, R., 'Wind Resistance of Merchant Ships', *The Royal Institution of Naval Architects. 115,* 1973.
22. VAN BERLEKOM, W., 'Wind Forces on Modern Ship Forms-Effects on Performance', *Trans. NECIES, 97,* 1981.
23. BLENDERMANN W., 'Parameter identification of wind loads on ships', *Journal of Wind Engineering & Industrial Aerodynamics*, 1994.
24. TOWNSIN, R. L., KWON, Y. J., BAREE, M. S., KIM, D. Y., 'Estimating the influence of weather on ship performance', *RINA, 135 - 191 – 209*, 1993.
25. ITTC, 7.5-04-01-01.2 'Recommended Procedures and Guidelines - Analysis of Speed / Power Trial Data', 2014.
26. FUJIWARA, T., UENO, M., IKEDA, Y., 'A New Estimation Method of Wind Forces and Moments acting on Ships on the basis of Physical Component Models', *JASNAOE 2*, 2005.
27. FUJIWARA, T., UENO, M., IKEDA, Y., 'Cruising performance of ship with large superstructures in heavy sea - 1st report: added resistance induced by wind', *Journal of the Japan Society of Naval Architects and Ocean Engineers. 2, 257-269*, 2006a.
28. HASSELMANN, K., et al. 'Measurements of Win-Wave Growth and Swell Decay During the Joint North

Sea Wave Project (JONSWAP)', *Deutschen Hydrographischen Institut, 12*, Hamburg, 1973.
29. PIERSON W.J., MOSKOWITZ, L.,'A Proposed Spectral Form for Fully Developed Wind Seas Based on Similarity Theory of S.S. Kitaigorodskii', *Journal Geoph. Res., 69, 5181 – 5190,* 1964.
30. DET NORSKE VERITAS, 'Environmental Conditions and Environmental loads', 2010.

 

# DEVELOPMENT OF RESISTANCE CRITERIA METHOD FOR SHIP PERFORMANCE EVALUATION BY ONBOARD MONITORING DATA

**A Sakurada, N Sogihara, M Tsujimoto,** National Maritime Research Institute, Japan
**H Sato,** MTI Co., Ltd., Japan

## SUMMARY

A collaborative project for evaluation of ship performance in actual seas called "OCTARVIA" was launched in October 2017 with the participation of 25 companies in the Japanese maritime cluster. One research theme of the project is evaluation of ship performance in a calm sea using onboard monitoring data, since performance in a calm sea is the basis for evaluating ship performance in actual seas and for assessing the effects of fouling and aging on ships. In this paper, the authors introduce the Resistance Criteria Method (RCM) for evaluation of ship performance in a calm sea. The method applies the increase rate of resistance from a calm sea for filtering of environmental factors, and also applies the apparent slip ratio for reduction of deviations in the filtered data. The effectiveness of RCM is validated by a comparison between onboard monitoring data and performance simulations based on the performance obtained by RCM.

## NOMENCLATURE

| | |
|---|---|
| $a_n$ | Coefficient decided by least-square method (-) |
| $b_n$ | Coefficient decided by least-square method (-) |
| $C_{DPC}$ | Criterion of $D_{PC}$ in RCM (-) |
| $d_{nv}$ | Coefficient decided by least-square method (-) |
| $d_{norm}$ | Normal distance between curve and each data point (-) |
| $D_{PC}$ | Evaluation index in RCM (-) |
| $FOC$ | Fuel consumption (t/d) |
| $FOC_{ms}$ | Measured fuel consumption per time step (t/d) |
| $FOC_{sim}$ | Fuel consumption by simulation (t/d) |
| $H$ | Significant wave height (m) |
| $H_{lim}$ | Criterion of significant wave height (m) |
| $N$ | Number of data of voyage (-) |
| $N_{all}$ | Number of all data (-) |
| $N_E$ | Engine revolution (min$^{-1}$) |
| $N_{eval}$ | Number of evaluation data (-) |
| $N_{fil}$ | Number of filtered data (-) |
| $N_{fit}$ | Number of fitting data (-) |
| $n_{id}$ | Engine revolution corrected to that in calm sea (min$^{-1}$) |
| $P$ | Engine power (kW) |
| $P_{MCR}$ | Engine power of MCR (kW) |
| $P_P$ | Propeller pitch (m) |
| $R_{id}$ | Resistance in calm sea (N) |
| $R_{ms}$ | Resistance in actual seas (N) |
| $S_A$ | Apparent slip ratio (-) |
| $\overline{S}_A$ | Mean value of apparent slip ratio (-) |
| $\hat{S}_A$ | Normalized apparent slip ratio (-) |
| $T$ | Mean wave period |
| $U_w$ | Wind speed (m/s) |
| $U_{wlim}$ | Criterion of wind speed (m/s) |
| $V_{dfc}$ | Design ship speed (knot) |
| $V_{rep}$ | Equivalent ship speed through water for representative displacement (knot) |
| $V_S$ | Ship speed through water (knot) |
| $V_{Ssim}$ | Ship speed through water by simulation per step time (knot) |
| $V_{Sms}$ | Measured ship speed through water per time step (knot) |
| $V_{S2}$ | Ship speed through water (m/s) |
| $V_{voy}$ | Ship speed through water obtained by measurement (knot) |
| $\beta$ | Drift angle (degree) |
| $\delta$ | Rudder angle (degree) |
| $\delta FOC$ | Integrated difference of fuel consumption (-) |
| $\delta L_w$ | Integrated difference of ship speed (-) |
| $\delta N_{fit}$ | Ratio of number of fitting data to all data (-) |
| $\Delta V$ | Difference between ship speed through water and ship speed over ground (knot) |
| $\delta R$ | Increase rate of resistance (-) |
| $\Delta R$ | Increase of resistance due to winds and waves (N) |
| $\delta R_{eval}$ | Criterion of $\delta R$ for evaluation data (-) |
| $\delta R_{fit}$ | Criterion of $\delta R$ for fitting data (-) |
| $\gamma$ | Wind angle (degree) |
| $\theta$ | Primary wave direction (0 degree is defined as head wave direction) (degree) |
| $\sigma_{SAH}$ | Standard deviation of normalized apparent slip ratio |
| $\nabla_{des}$ | Displacement volume in design full condition (m$^3$) |
| $\nabla_{rep}$ | Representative displacement volume (m$^3$) |
| $\nabla_{voy}$ | Displacement volume of each voyage (m$^3$) |

## 1. INTRODUCTION

The social need to reduce greenhouse gases (GHG) is increasing in the shipping sector. At the same time, improvement of sensors and the spread of data loggers are proceeding, and use of onboard monitoring data to study and improve the ship shape and method of operation is becoming a general practice.

Against this a background, the *OCTARVIA* project was started in 2017 and concluded in October 2020. This collaborative project, which involved 25 companies in the Japanese maritime cluster, aimed to establish an index for transparent and objective evaluation of ship life cycle performance. As a result, "Life cycle fuel consumption" was developed [1] integrating the technical elements.

In this project, the authors were responsible for developing an estimation and evaluation method for ship performance in a calm sea using onboard monitoring data. ISO 19030, which is a standard for measurement of changes in hull and propeller performance, provides a method for estimating ship performance [2]. Although this document suggests an effective method for assessing the trends of fouling and aging, a more detailed analysis is required in order to compare performance between different ships, including sister ships, and to evaluate new energy saving devices. As one problem of the ISO method, the suggested filtering criterion for wind is a single value of 7.9 m/s, even though the impact of winds differs depending on the ship size. Moreover, the fact that ISO 19030 does not treat wave effects is also considered to be a problem.

Sakurada et al. [3] proposed an estimation and evaluation method called the Resistance Criteria Method (RCM) for performance in a calm sea, in which the increase rate of resistance from a calm sea is used for filtering. The increase rate treats the level of environmental impacts equally independent of ship size. Since RCM includes a procedure for evaluating the estimated performance curve, results with assured high quality are obtained. The details of RCM are explained in the following chapter.

In the proposed RCM, the criterion of the increase rate of resistance was not discussed in detail. Therefore, in this paper, the criterion is studied and generalized based on results obtained with many ships of different types.

This paper introduces the RCM evaluation method and the study of parameters, summarizes the results of a simulation for validation, and finally presents the conclusions regarding the effectiveness of the proposed method.

## 2. RCM (Resistance Criteria Method)

### 2.1 DESCRIPTION OF RCM

RCM [3] is a method for obtaining a performance curve expressing the relationship of ship speed, engine revolution and engine power using onboard monitoring data. The entire procedure including the RCM to estimate the performance curve is described below.

1) Preliminary data filtering

1-1) Steady state

Because the analysis assumes the steady state in order to use mean data per time step (e.g., 60 min), filtering is carried out using the criteria in Table 1 in order to remove data in an unsteady condition.

1-2) Displacement

Because different displacements affect performance, only data of voyages in which the difference of displacement from the representative value is less than 5 % are used.

Table 1 Criteria of preliminary data filtering

| Item | Unit | Criteria | Adoption |
|---|---|---|---|
| $N_E$ | min $^{-1}$ | 40 % MCR | Upper |
| $\lvert\beta\rvert$ | degree | 3.0 | Lower |
| $\lvert\delta\rvert$ | degree | 5.0 | Lower |
| $\lvert\Delta V\rvert$ | knot | 0.5 | Lower |

Where $N_E$: engine revolution, $\beta$: drift angle, $\delta$: rudder angle, $\Delta V$: difference between ship speed through water and ship speed over ground.

2) Data correction

2-1) Displacement

To deal with differences in displacement, the ship speed through water is corrected using the admiralty coefficient shown in equation (1).

$$V_{rep} = V_{voy}\left(\frac{\nabla_{voy}}{\nabla_{rep}}\right)^{\frac{2}{9}} \tag{1}$$

Where $V_{rep}$: equivalent ship speed through water for the representative displacement, $V_{voy}$: ship speed through water obtained by measurement, $\nabla_{rep}$ : representative displacement volume, $\nabla_{voy}$ : displacement volume of each voyage.

2-2) Environmental factors

The effects of winds and waves are considered. Engine revolution and engine power are corrected to the ideal value in a calm sea by the Resistance-Thrust Identify Method [4]. Deviating data may be included due to the measurement accuracy of the speed log or anemometer, or the estimation accuracy of the wave height, wave angle and mean wave period, or the input accuracy of items relating to the ship shape.

3) RCM

3-1) Filtering by apparent slip ratio

To eliminate outliers, the apparent slip ratio expressed in equation (2) is introduced. The apparent slip ratio is normalized by equation (3), and is used for filtering by equation (4), in which the standard deviation of the normalized apparent slip ratio is used. The value of $C$ in equation (4) is studied in terms of the data distribution and number of remaining data, and was determined as $C$=1.0 [5].

$$S_A = 1 - \frac{V_{S2}}{P_p n_{id}} \tag{2}$$

$$\hat{S}_A = \frac{S_A - \bar{S}_A}{\bar{S}_A} \tag{3}$$

$$\left|\hat{S}_A\right| \leq C \cdot \sigma_{SAH} \tag{4}$$

Where $S_A$: apparent slip ratio, $V_{S2}$: ship speed through water (m/s), $P_P$: propeller pitch, $n_{id}$: engine revolution corrected to that in a calm sea, $\hat{S}_A$: normalized apparent slip ratio, $\bar{S}_A$: mean value of apparent slip ratio, $\sigma_{SAH}$: standard deviation of normalized apparent slip ratio.

3-2) Filtering by increase rate of resistance

In the procedure of data correction for environmental factors such as winds and waves, the increase rate of resistance from a calm sea is calculated for each of the data expressed by equations (5) and (6).

$$\delta R = \frac{\Delta R}{R_{id}} \tag{5}$$

$$R_{id} = R_{ms} - \Delta R \tag{6}$$

Where $\delta R$: increase rate of resistance, $\Delta R$: increase of resistance due to winds and waves, $R_{id}$: resistance in a calm sea, $R_{ms}$: resistance in actual seas.

RCM includes two types of filtering, one for fitting and the other for evaluation of the fitting curve, as this is better for maintaining the number and range of data for fitting. However, for high quality estimation, it is preferrable to extract data which include less error related to correction for environmental factors. Because these two types of filtering conflict when determining the criteria of $\delta R$, $\delta R_{fit}$ and $\delta R_{eval}$ are used for the fitting criterion and for the evaluation criterion, respectively.

Using the data filtered by $\delta R_{fit}$, which has a comparatively large value, the performance curve is expressed by the least-square method in equation (7) and equation (8). The data used here are called "fitting data".

$$N_E = d_{nv} \cdot V_S \tag{7}$$

$$P = a_n \cdot N_E^{\,b_n} \tag{8}$$

Where $P$: engine power, $a_n$, $b_n$, $d_{nv}$: coefficients decided by the least-square method.

$\delta R_{eval}$, which has a comparatively small value, is used in the following evaluation procedure.

3-3) Evaluation of estimated curve

Now the performance curve is obtained. In RCM, the curve is treated as a temporary one, and is evaluated by the evaluation data filtered by $\delta R_{eval}$.

First, an evaluation index $D_{PC}$ is calculated by equation (9).

$$D_{PC} = \sqrt{\frac{\sum_{i=1}^{N_{eval}} \{d_{norm}(i)\}^2}{N_{eval}}} \tag{9}$$

Where $N_{eval}$: number of evaluation data, $d_{norm}(i)$: normal distance between curve and each of the evaluation data.

The normal distance $d_{norm}$ is obtained from the normalized relationship of ship speed and engine power divided by the design ship speed $V_{dfc}$ and the engine power of MCR $P_{MCR}$.

In the *OCTARVIA* project, the target differences between the simulated and measured ship speed and fuel consumption were set as 2.0 % in both cases. Based on the idea that the difference of the performance curve should be less than 2.0 %, it is appropriate that the criterion of $D_{PC}$ ($C_{DPC}$) is 2.0 %.

If the calculated result of $D_{PC}$ is smaller than $C_{DPC}$, the curve is assured as having high quality. If $D_{PC}$ is larger than $C_{DPC}$, the fitting procedure is repeated by reducing $\delta R_{fit}$, and a high quality curve is obtained after the iteration.

The simplified flow of RCM is shown in Figure 1.

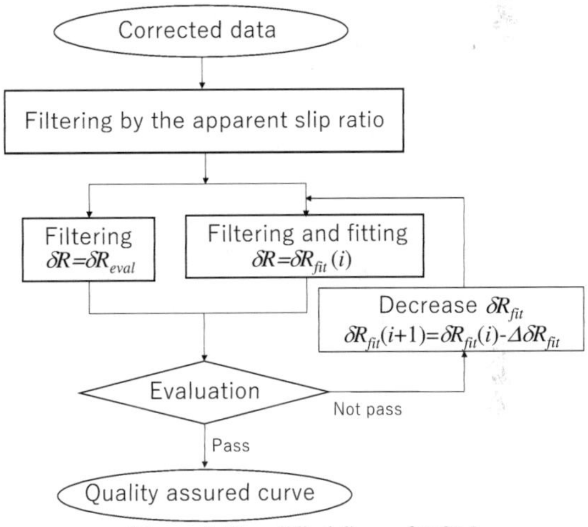


Figure 1 Simplified flow of RCM

Validation has been already done using $\delta R_{eval}$ of 2 % and $\delta R_{fit}$ of 100 % [6]. However, whether the values are reasonable or not has not been discussed. Therefore, the next chapter explains the study of the reasonable values of $\delta R_{eval}$ and $\delta R_{fit}$.

## 2.2 CRITERIA OF ENVIRONMENTAL FACTORS

For the study of $\delta R_{eval}$ and the initial $\delta R_{fit}$ in the iteration, the number of remaining data and the evaluation index $D_{PC}$ are studied based on the criterion $\delta R$ in Table 2. This analysis is conducted after processes 1) to 3-1) in 2.1.

The ships used in the study are listed in Table 3. The representative displacements were decided considering the number of data. As examples, Figure 2 and Figure 3 show the number of remaining data and $D_{PC}$ of the container ships CS1 and CS2. $N_{fil}$ is the number of remaining data, and $N_{case1}$ is the number of remaining data in Case 1. These calculations are conducted to decide the appropriate initial value of $\delta R_{fit}$.

Table 2 Filtering criteria

| Case | $\delta R$ (%) | |
|---|---|---|
| | Min | Max |
| 1 | 0 | 100 |
| 2 | 0 | 50 |
| 3 | 0 | 30 |
| 4 | 0 | 20 |
| 5 | 0 | 10 |
| 6 | 0 | 5 |
| 7 | 0 | 3 |

Table 3 Ships used in study

| ID | Ship type | Condition |
|---|---|---|
| CS1 | Container ship | 1 (Laden) |
| CS2 | Container ship | 1 (Laden) |
| PC1 | PCC | 1 (Laden) |
| PC2 | PCC | 1 (Laden) |
| BC1 | Bulk carrier | 2 (Laden), 3 (Ballast) |
| BC2 | Bulk carrier | 1 (Laden) |
| TA1 | Tanker | 2 (Laden/Ballast) |

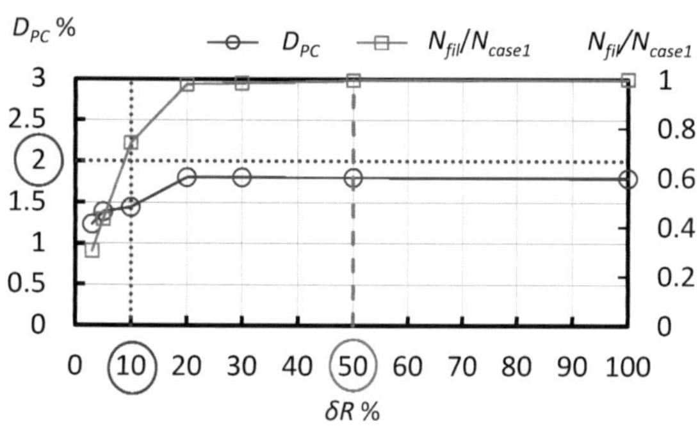

Figure 2 Number of data and evaluation index (CS1)

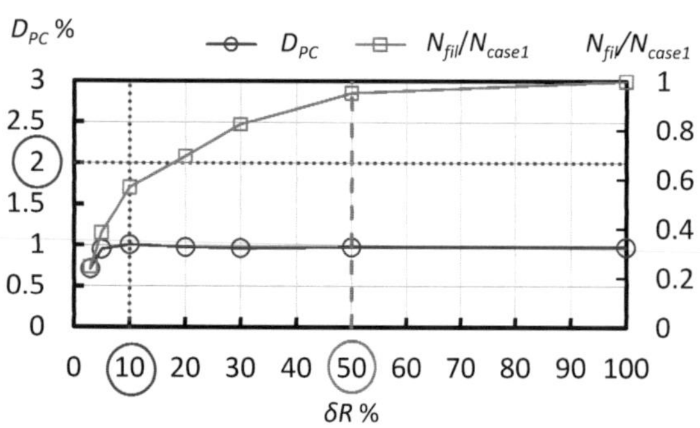

Figure 3 Number of data and evaluation index (CS2)

For all ships and displacements, more than 90 % of the data remain at $\delta R$=50 %, as shown in Figure 2 and Figure 3, and in some cases, the number of data decreases from $\delta R$=50 %, as in Figure 3. Thus, this study shows that it is appropriate to use 50 % as the initial value of $\delta R_{fit}$.

$D_{PC}$ must be smaller than $C_{DPC}$ (2.0 %) at $\delta R=\delta R_{eval}$ in order to pass the iteration. For most of the ships and displacements, $D_{PC}$ is less than 2.0 % at $\delta R$=10 % and becomes smaller as $\delta R$ approaches 0, as shown in Figure 2 and Figure 3. However, Figure 4 shows a different trend of $D_{PC}$ at 3 displacements for the ballast condition of the bulk carrier BC1. In that figure, BC1_B1, BC1_B2 and BC1_B3 mean the 1st, 2nd and 3rd displacement groups of BC1, respectively.

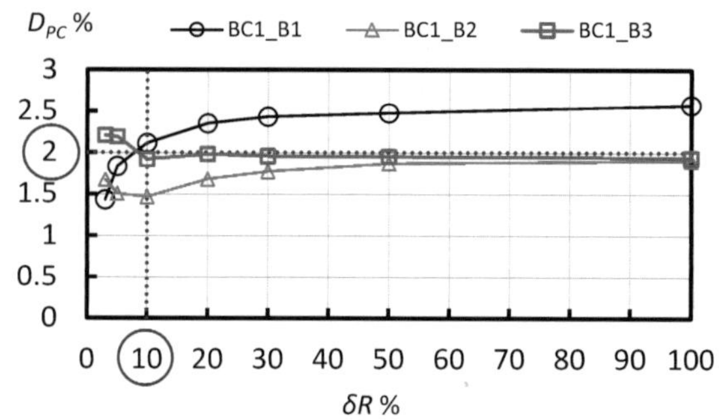

Figure 4 Evaluation index $D_{PC}$ (BC1 Ballast)

The $D_{PC}$ values of BC1_B2 and BC1_B3 have their minimum values at $\delta R$ =10 %. This means that filtering of $\delta R$ is meaningless for eliminating scattered data in these cases if $\delta R$ is smaller than 10 %.These considerations show that $\delta R_{eval}$=10 % is reasonable.

In cases like BC1_B1, in which $D_{PC}$ exceeds $C_{DPC}$ at $\delta R_{eval}$, the curve cannot pass the iteration. In such cases, the final curve is obtained by the initial value of $\delta R_{fit}$ and is qualified as having poor accuracy. The quality of performance curves is discussed in detail in paper [6].

## 3. VALIDATION

A simulation of the ship speed and fuel consumption was conducted for validation of the proposed evaluation method. The flow of the validation was as follows.

1) Estimation of the performance curve in a calm sea using onboard monitoring data.

2) Simulation of the ship speed and fuel consumption using the performance curve estimated above.

3) Comparison of the ship speed and fuel consumption in the simulation result and onboard monitoring data.

The target ship is a medium range tanker whose principal particulars are shown in Table 4.

Table 4 Principal particulars of medium range tanker

| Ship type | Tanker |
|---|---|
| Length between perpendiculars | 185.0 m |
| Breadth | 32.2 m |
| Design draft | 13.0 m |

### 3.1 ESTIMATION OF PERFORMANCE CURVE

The three methods in Table 5 were used in estimation of the performance curve. M-1 and M-2 are assumed as conventional methods, and M-3 is RCM. The performance model expressed by equations (7) and (8) is applied to all three methods.
The data filtered by a 5 % difference from the representative displacement were included in the analysis, and the ship speed was corrected by equation (1).
Two displacements were chosen as representative values, as shown in Table 6, where $\nabla_{des}$ is the displacement volume in design full condition. The groups are named G1 and G2. G1 includes 3 voyages (v1, v3, v7), and G2 includes 5 voyages (v4, v6, v8, v10, v12).

Table 5 Filtering criteria

| | M-1 | M-2 | M-3 |
|---|---|---|---|
| Performance model | Equations (7), (8) | | |
| Data | Group of voyages (G1, G2) | | |
| Filtering (steady state) | Filtered | | |
| Correction (environmental factors) | Not corrected | | Corrected |
| Filtering (environmental factors) | $U_w \le U_{wlim}$ | $U_w \le U_{wlim}$<br>$H \le H_{lim}$ | RCM<br>$\delta R_{eval}$=10 %,<br>initial value of $\delta R_{fit}$=50 % |

Table 6 Representative displacement volumes

| Group ID | Displacement volume |
|---|---|
| G1 | 80% $\nabla_{des}$ |
| G2 | 51% $\nabla_{des}$ |

Correction for environmental factors was not applied in M-1 and M-2, as these are conventional methods, and was applied to the simulation conducted by M-3. Filtering for environmental factors was carried out for wind speed ($U_w$) in M-1, for wind speed ($U_w$) and significant wave height ($H$) in M-2 and for $\delta R$ in M-3.
The criterion of wind speed $U_{wlim}$ was 7.9 m/s, and that of significant wave height $H_{lim}$ (meter) was expressed by equation (9), assuming that the impact is different due to the ship size.

$$H_{lim} = 1.35\sqrt{\frac{L_{PP}}{100}} \tag{9}$$

Where $L_{PP}$: length between perpendiculars (m). The data are 1 hour averaged values. The data used in the analysis were acquired within one year after the ship was put into service so as not to be affected by fouling or aging of the ship. The wind speed and wind angle were obtained by the anemometer on the ship.

Waves require particular attention, as the spectral shape is important for estimating the added resistance in waves in correction for environmental factors and simulations of ship speed and fuel consumption. The standard spectrum is used in the calculation. The spectrum is expressed by the separation form of the frequency spectrum and angular distribution function.
Because the wave is assumed to be a wind wave, the frequency spectrum is expressed by the modified Pierson-Moskowitz spectrum [7] with the IACS spectrum [8]. The angular distribution function is expressed as the second power of the cosine, and the significant wave height, wave period and wave angle are hindcast data provided by the Japan Weather Association [9].
The results of the estimation of the performance curve and the data used in G1 are shown in Figure 5, where "ms" means measured data, "fit" and "eval" are the fitting data and the evaluation data in RCM, respectively, and FIT is the performance curve by each method. The methods are indicated in parentheses.

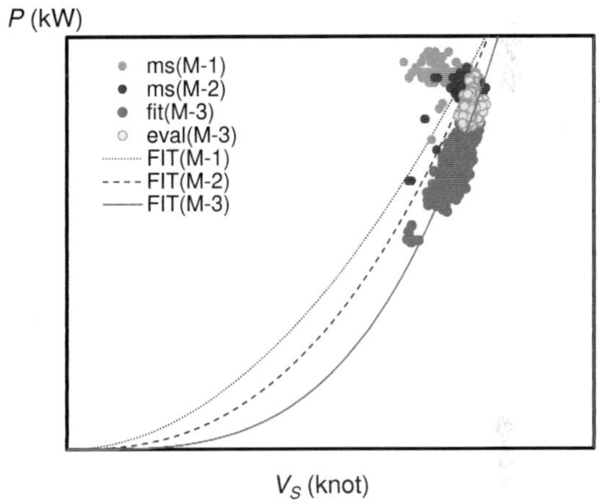


Figure 5 Data and performance curves (G1)

The engine power at the same ship speed is lowest in FIT(M-3), second lowest in FIT(M-2), and highest in FIT(M-1). This difference occurs because M-1 includes a large environmental effect, while M-3 includes a smaller environmental effect than the other two methods.
Figure 6 shows the evaluation index $D_{PC}$ and the remaining data in G1 of each method.
In the calculation of $D_{PC}$ by M-1 and M-2, the fitting data and the evaluation data are the same in both methods, whereas the fitting data and evaluation data are defined by RCM in M-3. In the figure, $\delta N_{fit}$ is used for the number of the data obtained by equation (10).

$$\delta N_{fit} = \frac{N_{fit}}{N_{all}} \tag{10}$$

Where $N_{fit}$: number of data used for fitting, $N_{all}$: number of all data.

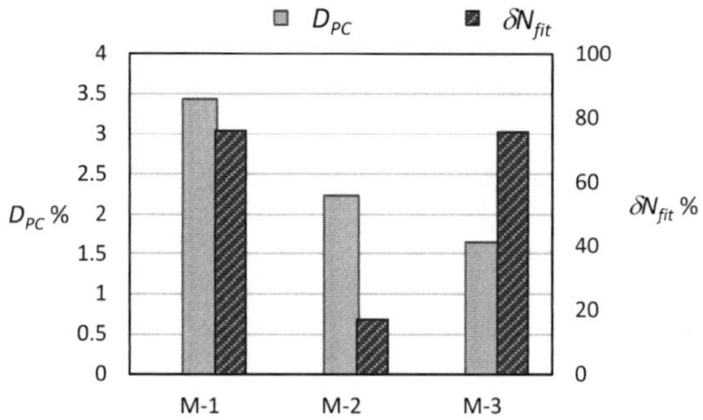


Figure 6 Evaluation index and remaining data (G1)

From Figure 6, a sufficient number of data remains in M-1, but because $D_{PC}$ is large, the data are scattered from the fitting line. In M-2, the value of $D_{PC}$ is improved compared with M-1, but the number of data is small. On the other hand, in M-3, the number of remaining data is similar to that in M-1 and $D_{PC}$ has the smallest value. This means that the data used for fitting cover includes a sufficient number of data, and the reliability of the data is assured.

## 3.2 SIMULATION RESULTS

The simulation of ship speed and fuel consumption was carried out by VESTA [10] using the estimated performance curve, in which the ship speed was corrected to the equivalent value for the displacement of each voyage by equation (1). In the calculation, added resistance in waves, added resistance due to wind, and the added resistance due to drift angle and rudder angle are estimated and the solution is found by the equilibrium equation.

The coefficients relating to the ship shape, propeller open characteristics, thrust deduction and relative rotative efficiency were estimated by UNITAS [11] based on the principal particulars of the ship.

As one example of the simulation, the result for voyage 3 (G1) is shown from here. Figure 7 shows the weather conditions, and Figure 8 shows the engine revolution of the input.

Where $H$: significant wave height, $U_w$: wind speed, $\theta$: wave angle (0 deg. is defined as the head wave direction), $\gamma$: wind angle (0 deg. is defined as the head wind direction), $T$: mean wave period.

Figure 9 shows the ship speed through water ($V_S$) and fuel consumption ($FOC$) by M-1, M-2 and M-3, where "ms" means measured data and "sim" means simulation results.

In M-1, the simulation result of the ship speed and fuel consumption is 0 at the data around No. 470. Here, the significant wave height and wind speed are severe, and both the wave and wind angles are head. Based on these facts, the performance curve of the simulation by M-1 is higher than the actual one. Therefore, resistance in a calm sea is overestimated and there is no solution in the equation of equilibrium. The simulation result of ship speed is underestimated and that of fuel consumption is overestimated during the whole voyage. This is also caused by the performance curve, which includes a larger environmental effect.

In M-2, the solution is found at the data around No. 470. However, the simulation still underestimates ship speed and overestimates fuel consumption during the whole voyage.

In M-3, the difference between the measured data and the simulation result is small during the whole voyage.

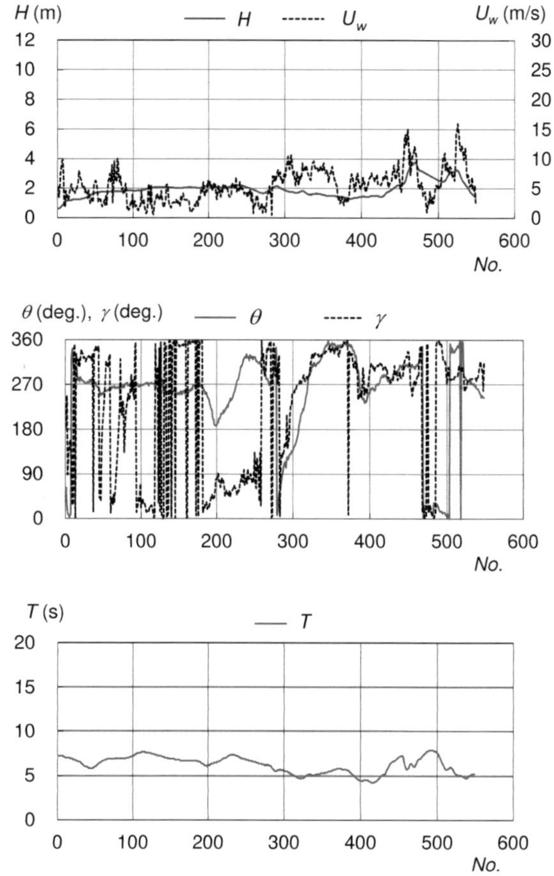


Figure 7 Weather conditions in voyage 3

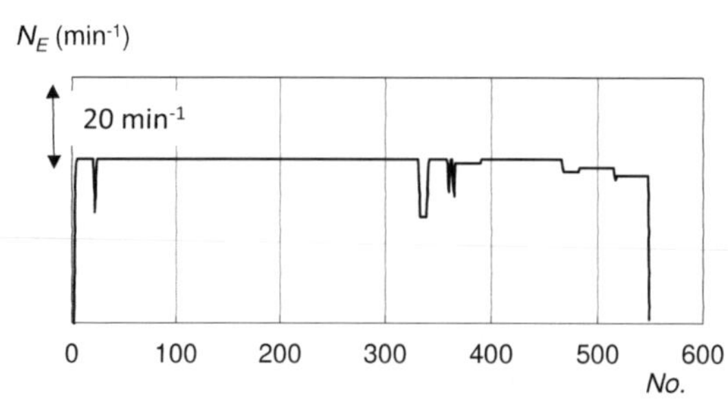


Figure 8 Engine revolution of input

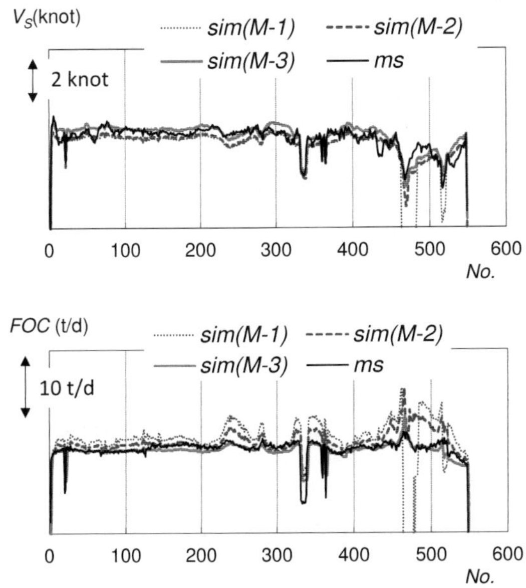


Figure 9 Simulation results

## 3.3 COMPARISON OF SIMULATION RESULTS AND ONBOARD MONITORING DATA

From this point, the calculation for all voyages will be explained. The differences of the integrated values of ship speed and fuel consumption are calculated by equation (11) and equation (12), respectively. It should be noted that the calculation by M-1 does not include the data which gave the simulation result of 0 around No. 470.

$$\delta L_W = \frac{\sum_{i=1}^{N} V_{Ssim}(i) - \sum_{i=1}^{N} V_{Sms}(i)}{\sum_{i=1}^{N} V_{Sms}(i)} \tag{11}$$

$$\delta FOC = \frac{\sum_{i=1}^{N} FOC_{sim}(i) - \sum_{i=1}^{N} FOC_{ms}(i)}{\sum_{i=1}^{N} FOC_{ms}(i)} \tag{12}$$

Where $\delta L_W$: integrated difference of ship speed, $V_{Ssim}$: ship speed through water by simulation per time step, $V_{Sms}$: measured ship speed through water per time step, $N$: number of data of voyage, $\delta FOC$: integrated difference of fuel consumption, $FOC_{sim}$: fuel consumption by simulation per time step, $FOC_{ms}$: measured fuel consumption per time step.

The differences in ship speed and fuel consumption by the three methods are shown in Figure 10. Improved simulation results for the ship speed through water and fuel consumption were obtained with M-3 for almost all voyages. The red broken lines in the figure indicate 2.0 % and -2.0 %. With one slight exception, the results by M-3 are within ±2.0% for both the ship speed and fuel consumption.

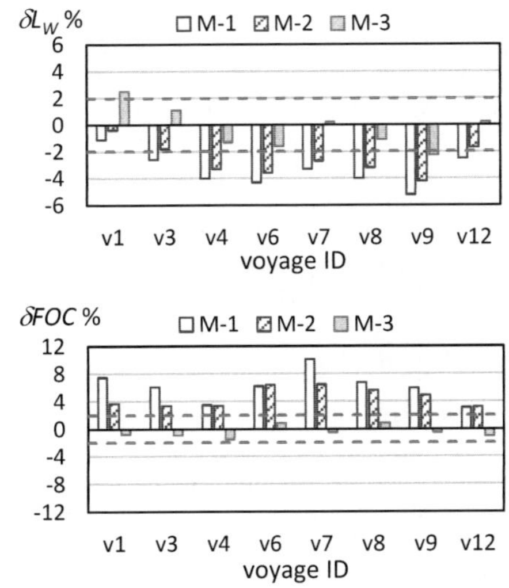


Figure 10 Difference of ship speed and fuel consumption (all voyages)

The mean value and standard deviation of $\delta L_W$ and $\delta FOC$ of all voyages are shown in Table 7 and Table 8, respectively.

Table 7 Mean/standard deviation for all voyages ($\delta L_w$)

| Method | M-1 | M-2 | M-3 |
|---|---|---|---|
| Mean % | -3.4 | -2.6 | -0.3 |
| Standard deviation % | 1.2 | 1.2 | 1.5 |

Table 8 Mean/standard deviation for all voyages ($\delta FOC$)

| Method | M-1 | M-2 | M-3 |
|---|---|---|---|
| Mean % | 6.1 | 4.6 | -0.4 |
| Standard deviation % | 2.1 | 1.3 | 0.8 |

The mean values of $\delta L_W$ and $\delta FOC$ approach 0 in the order of M-1, M-2 and M-3. From these results, it is obvious that M-3 is more effective than the conventional methods M-1 and M-2.

## 4. CONCLUSIONS

This paper introduced the Resistance Criteria Method (RCM) evaluation method for obtaining performance curves in a calm sea by using onboard monitoring data. RCM can evaluate performance curves with high accuracy by filtering by the apparent slip ratio and the increase rate of resistance. In RCM, filtering by the increase rate of resistance is carried out by two methods in parallel. One is used to obtain the performance curve, and the other, to evaluate the obtained curve. The data used for fitting are called fitting data, while those used for the evaluation are called evaluation data.

The reasonable criterion for the increase rate of resistance was investigated by using 7 ships of various ship types. The investigation revealed that the reasonable initial criterion of the fitting data is 50 % of the increase rate of resistance from a calm sea, and the criterion for the evaluation data is 10 % of the increase rate.
Based on these criteria, a validation was conducted by simulation of the ship speed and fuel consumption in 8 voyages by a medium range tanker. To compare the methods, the integrated values of the ship speed and fuel consumption were calculated for all voyages. When using the proposed method, the difference in the integrated values of ship speed and fuel consumption were within ±2.0 % for almost all voyages. Compared with the conventional methods, the effectiveness of the proposed method is clear.

## 5. ACKNOWLEDGEMENTS

This research was carried out by the Project for Evaluation of Ship Performance in Actual Seas – Japan Maritime Cluster Collaborative Research – (OCTARVIA Project). We would like to express our deepest gratitude to all the parties concerned who discussed the content of this research in the working groups.

## 6. REFERENCES

1. Tsujimoto, M., Sogihara, N., Sato, H., Kume, K., Orihara, H., Sugimoto, Y., Kuroda, M., 'Development of Performance Evaluation of Ships in Actual Seas – OCTARVIA Project -', *Proc. of HullPIC 2020*, 2020.
2. ISO, 'Ship and marine technology - Measurement of changes in hull and propeller performance- Part 3: Alternative methods', *ISO 19030-3*, 2016.
3. Sakurada, A., Sogihara, N., Tsujimoto, M., 'Development of a Filtering Method for Evaluation of Performance in a Calm Sea Based on Onboard Monitoring Data', *Journal of the Japan Society of Naval Architects and Ocean Engineering, Vol. 31*, 2020
4. ITTC, 'Prediction of Power Increase in Irregular Waves from Model Test (RP.7.5.-02-07-02.2)', *International Towing Tank Conference*, 2014
5. Sogihara, N., Sakurada, A., Tsujimoto, M., 'Validation of Filtering Method for Evaluating Ship Performance in Calm Sea Using Onboard Monitoring Data', *Journal of the Japan Society of Naval Architects and Ocean Engineering* (in submission)
6. Sogihara, N., Sakurada, A., Tsujimoto, M., Sato, H., 'Validation of the Evaluation Method of Ship Performance Using Onboard Monitoring Data – Development of Resistance Criteria Method with Apparent Slip Ratio and Quality Management -', *Proc. of HullPIC 2020*, 2020.
7. Pierson, W. J. Jr., Moskowitz, L., 'A Proposed Spectral Form for Fully Developed Wind Seas Based on the Similarity Theory of S. A. Kitaigorodskii', *Journal of Geophysical Research, Vol. 69*, 1964.
8. International Association of Classification Societies, 'Standard Wave Data', *IACS Rec. No.34,* Rec.2000/Corr.2001.
9. Sato, Y., Matsuura, K., 'Forecasting and Hindcasting', *The Naval Architect, November 2019,* 2019.
10. Tsujimoto, M., Sogihara, N., Kuroda, M., Sakurada, A., 'Development of a Ship Performance Simulator in Actual Seas', *Proc. of OMAE2015,* 2015.
11. Sogihara, N., Kuroda, M., Sakurada, A., Yokota, S., 'UNITAS: Tool for Supporting Evaluation of Ship Performance in Actual Seas', *Papers of National Maritime Research Institute Vol. 19, No. 1*, 2019

## 7. AUTHOR BIOGRAPHIES

**Akiko Sakurada** works at the National Maritime Research Institute, MPAT, Japan, as a Researcher in the Performance of Ships in Actual Seas Research Group, Fluids Engineering & Hull Design Department. She is engaged in analysis of ship monitoring data, estimation of fuel consumption in actual seas and evaluation of the effect of weather on ships operating in actual seas.

**Naoto Sogihara** works at the National Maritime Research Institute, MPAT, Japan, as a Chief Researcher in the Performance of Ships in Actual Seas Research Group, Fluids Engineering & Hull Design Department. He is responsible for the evaluation method for ship performance in actual seas with data obtained by full-scale measurements.

**Masaru Tsujimoto**, Ph.D. MRINA, and Director, Fluids Engineering & Hull Design Department, National Maritime Research Institute, MPAT, Japan. He is an expert on ship performance in actual seas. He contributes to the development of new hull forms that reduce added resistance in waves and the weather database and advanced weather routing system.

**Hidehiko Sato** works at MTI Co., Ltd that is a R&D firm under NYK Line, as General Manager of Maritime & Logistics Technology Group. He contributes to the evaluation method for ship performance in actual seas with data obtained by full-scale measurements.

# JOINT SEA TRIAL OF SHIPS WITH GATE RUDDER® AND CONVENTINAL RUDDER

**N Sasaki**, University of Strathclyde, UK
**S Kuribayshi**, Kuribayashi Steamship Co., Japan
**N Bulten,** Wärtsilä, The Netherlands
**M Yazawa,** Kamome Propeller Co.,Ltd, Japan

## SUMMARY

This paper introduces full scale performance results obtained from a joint sea trial of two sister ships. Shigenobu equipped with Gate Rudder® was delivered Dec. 2017 and sister ship Sakura equipped with a conventional flap rudder was delivered Aug. 2016 from Yamanaka Shipyard. Both ships are exactly the same ship except for their rudder systems. According to the shipyard, standard sea trials of the two ships were conducted and 14% of power/fuel saving was found for Gate Rudder®. After 30 months from delivery of Shigenobu, the voyage data from both vessels revealed that the gain could be much larger: 33%. In order to investigate the mechanism of large energy saving during the navigation period and eliminate all the uncertainty effected by the place and weather, the joint sea trial was planned and put into action at the end of 2020.

In this paper, the full-scale operational performance and manoeuvring differences obtained from the joint sea trial and analysis supported by CFD are presented.

## NOMENCLATURE

| | |
|---|---|
| $C_T$ | Propeller loading factor (-) |
| $H_W$ | Significant Wave Height (m) |
| $K_q$ | Torque coefficient |
| M | Vessel mass (kg) |
| $P$ | Power measured in service (kW) |
| $t$ | Thrust deduction factor (-) |
| $T_P$ | Propeller thrust (N) |
| $V_W$ | Relative wind velocity (m/s) |
| $V_c$ | Current velocity (m/s) |
| $V_{GPS}$ | GPS based vessel velocity (m/s) |
| $V_{LOG}$ | Log based vessel velocity (m/s) |
| $V_s$ | Vessel speed (m/s) |
| $w$ | Wake factor (-) |
| $\alpha$ | Added mass factor |
| $\psi$ | Relative wind direction (deg.) |

Figure 1 HMS Rattler and HMS Alecto in their 1845 competition

## 1 INTRODUCTION

The ship performance after delivery of the vessel is not easy to evaluate because of a lot of uncertainties such as the effect of wind, waves and current in addition to the ageing effect. If two vessels were designed based on the same hull form and the same propulsion system except their rudder systems as shown in

Photo 1, these two ships can show a clear difference of performance due to these rudder systems. However, it is necessary to conduct such joint trials where the weather conditions are very close for both ships. The first idea of this trial is the bollard pull battle of the Rattler, the most famous competition in marine propeller history.

It is also notable fact that the rudder has not been changed and remained as it was installed behind the propeller while the paddle was replaced by a new comer (propeller).

Photo 1 Flap Rudder (upper) Gate Rudder (lower).

 

Table 1 Principal dimensions of Sakura (Flap Rudder) & Shigenobu (Gate Rudder).

| | Sakura | Shigenobu |
|---|---|---|
| Loa (m) | 111.4 | |
| B (m) | 17.8 | |
| d (m) | 5.24 | |
| Main Engine | 3309kW x 220rpm | |
| Rudder | Flap Rudder | Gate Rudder |
| Delivery | August 2016 | December 2017 |

Photo 1 and Table 1 show a comparison of the conventional rudder and the Gate Rudder® system on the two sister vessels which are the subject of this paper. The synchronised control of the two rudder blades also provides effective control of the propeller slipstream and hence steering, thus the Gate Rudder® system presents not only more propulsive efficiency but also greater manoeuvrability. In addition to these two major advantages there are other performance improvements based on further analysis of the voyage data, including reduced resistance during seakeeping as well as manoeuvring motion. Such preferred full-scale features are explained in section 3 and 4.

## 2 JOINT SEA TRIAL

### 2.1 Background

After delivery of both ships, the differences in their propulsive performance through their voyage data was huge and more than 30% so far. Figure 2 shows the differences in fuel consumption obtained from the voyage data of the two sister ships. Each point represents one voyage from the departure port to the arrival port. Both ships were running the same route at some period as shown in the Photo 2.

The difference in fuel consumption is very large, being 19% for the West bound route and 35% for North bound route where the weather is more severe than for the West bound route.

However, we need to investigate the reason why we can see so large difference due to the rudder systems only.

One day, we noticed that two sister ships would have a chance to cross near Kobe which is the mother port of both ships. We planned this joint sea trial at the beginning of 2020 to occur in May. However, we postponed this trial due to Covid '19.

### 2.1 Hearing from the captain

The captain of Shigenobu has unique experience with both ships 'Sakura' and 'Shigenobu'. He was a captain for two ships continuously. He operated Sakura before Dec. 2017 when Shigenobu was delivered. From this reason, he knows both ships very well. He compared Sakura and Shigenobu except for narrow channel operation because he had no experience with this operation of Sakura. Table 2 shows the summary of his evaluation according to his experience with two ships. Unfortunately, we have no other captain who has operated both ships, which limits the review to this single experienced captain.

Table 2 The captain's evaluation with two ships.

| | Sakura | Shigenobu |
|---|---|---|
| Fuel Saving | ★★★ | ★★★★★ |
| Couse keeping | ★★★ | ★★★★ |
| Circle (High S) | ★★★ | ★★★ |
| Circle (Low S) | ★★★ | ★★★★★ |
| Seakeeping | ★★★ | ★★★★★ |
| Noise | ★★★ | ★★★★ |
| Rolling | ★★★ | ★★★★ |
| Running Anchor | ★★★ | ★★★★★ |

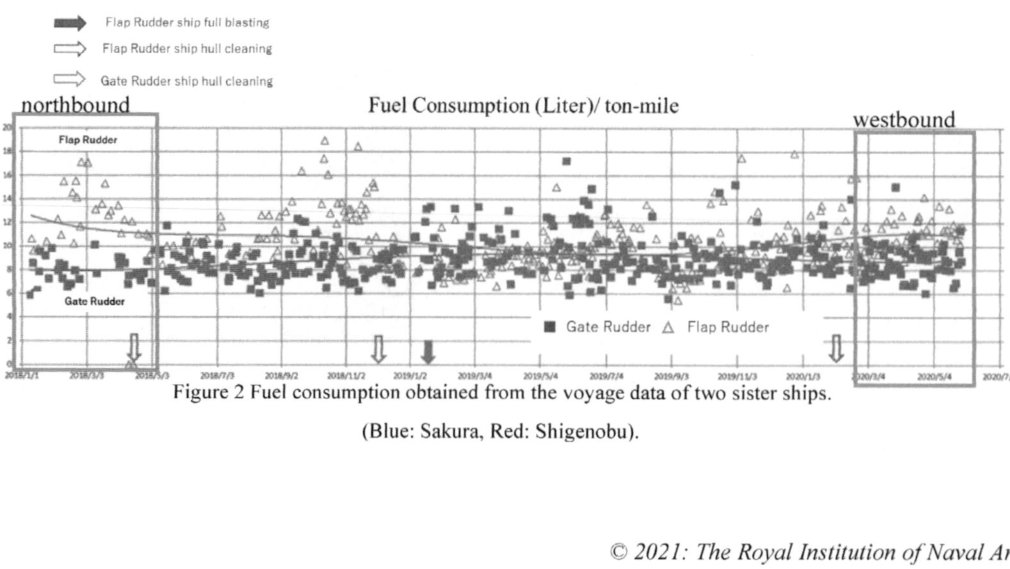


Figure 2 Fuel consumption obtained from the voyage data of two sister ships.

(Blue: Sakura, Red: Shigenobu).

He mentioned that remarkable differences are as follows:

(1) Shigenobu is faster than Sakura abt. 1kts
(2) Shigenobu is very tough in a storm
(3) Anchor of Shigenobu is more effective than Sakura's

## 2.2 Test Conditions

The test conditions were selected to satisfy two important requirements. One is that the condition is close to the sea trial conditions by which we obtained 14% power difference by their speed trials. Second is the condition by which we can see the difference of their course keeping ability. A discussion was made on the second point. The opinion from the Sakura side, was that they preferred the normal condition for which both ships are running usually. The opinion from the Shigenobu side, was that they preferred the critical condition where the difference between both ships is clearly detected. The final decision was made based on the latter opinion as shown in Figure 3.

From the captains of Sakura and Shigenobu, it was interesting to know that the captain of Sakura regarded the larger trim as important while the captain of Shigenobu focused on the smaller trim to avoid the damage of containers during the loading/unloading period. The captain of Sakura explained that he took the better course keeping ability seriously.

From the tank test result, smaller trim showed better efficiency. Therefore, it was expected that Sakura would show slightly better efficiency compared with obtained voyage data.

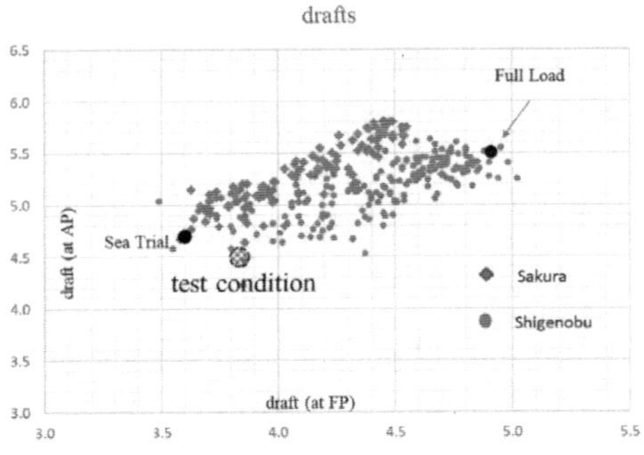


Figure 3 Operational draft conditions and trial drafts

## 2.3 Test Area

The test area was selected as shown in Figure 4 where both ships can sail within a few hours from their mother port Kobe and many shipyards have been using this area as a sea trials course. In order to avoid any risks which may occur, information was collected from the local shipyard who knows this area very well.

As the north part of this area is close to the Akashi strait where the strong current exists, the tide chart was investigated and discussed about the best place. Luckily, the strong west wind on the day was protected by the island mountain and sea state is around BF 4 ($V_w$ = 5-6m/s, $H_w$ =0.8m) while sea state of other side of the Osaka gulf was terrible.

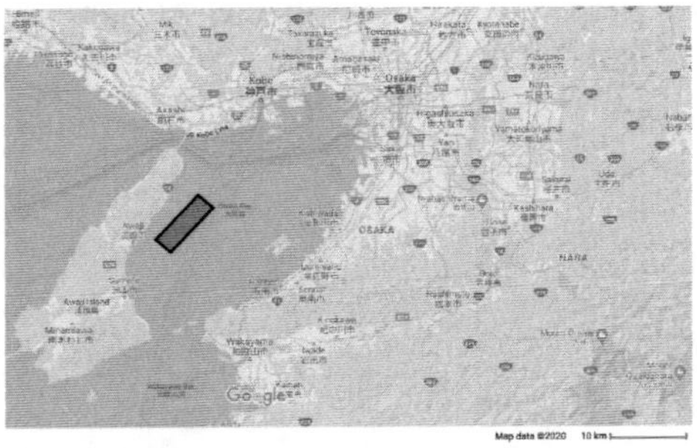

Figure 4 Test are for the joint seatrial

## 2.4 Speed trial

### 2.4.1 Results of speed trial

The speed trial was conducted by running the vessels closely with minimum time lags. The results are shown in Figure 5 and 6/

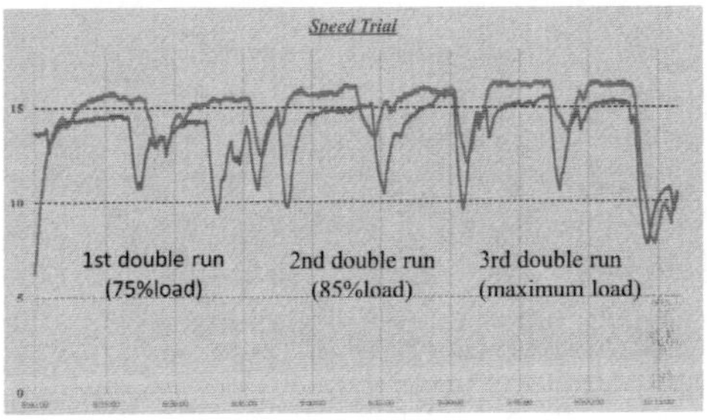


Figure 5 Ship speed vs time comparison from sea trial data (Red: Shigenobu, Blue: Sakura).

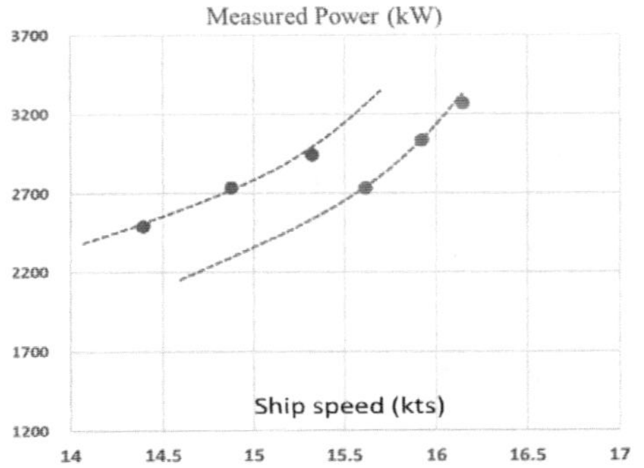


Figure 6 Measured Power against ship speed (Blue: Sakura, Red: Shigenobu).

The test condition was decided to keep the same rpm for both ships because they were using the same engine telegraph curves (rpm – power). In addition to this information, both ships have the power monitoring system based on the engine performance which seems more accurate.

The ship speeds were measured by two instruments, one a doppler log which is tuned already, the second a GPS located on the navigation bridge. Figure 7 shows the differences between both instruments and it show that the error is negligibly small compared with the difference of both ship speeds.

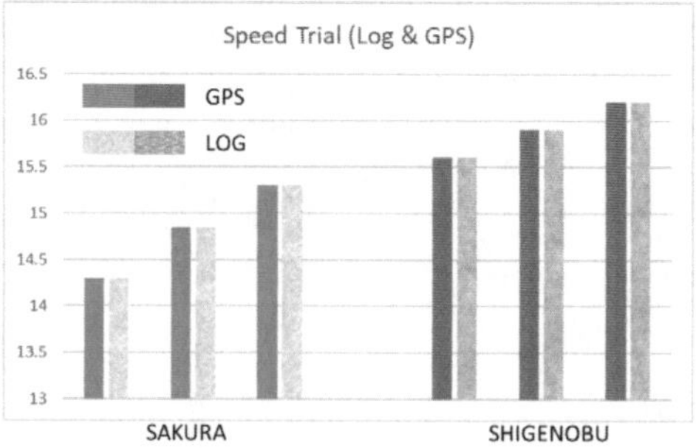


Figure 7 Comparison of the LOG and GPS data for both ships.

The trials were conducted on 30th Dec. 2020 and we sailed both ships from Kobe port as shown in Photo 2.

Photo 2 The test field near Awaji Island.

The accuracy can be checked by current which can be measured by following equation.

$$V_C(t) = V_{GPS}(t) - V_{LOG}(t) \qquad (1)$$

Where, $V_C$ , VGPS and VLOG is current , GPS and LOG speeds respectively.

The current speeds obtained from the equation (1) shows the same trend predicted by Japan Meteorological Agency as shown in Figure 8.

Table 3 Ship speeds at certain engine power

| | Sakura | Shigenobu | difference |
|---|---|---|---|
| 2500kW | 14.40 | 15.35 | 0.95kts |
| 2750kW | 14.90 | 15.65 | 0.75kts |
| 3000kW | 15.40 | 15.90 | 0.50kts |

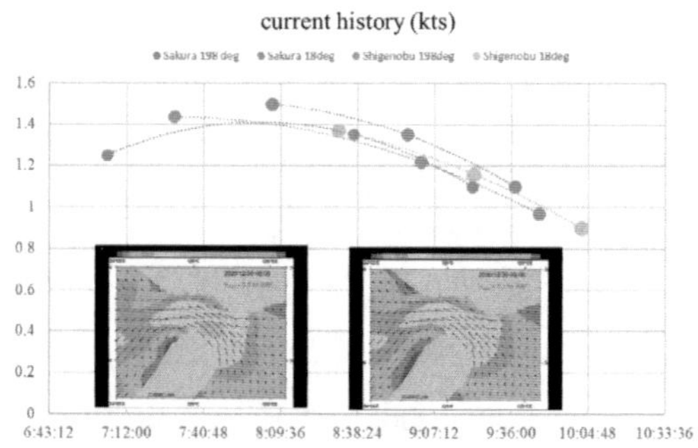


Figure 8 Current speed history against time.

It seems that the effect of wind and waves appeared more largely on the Shigenobu's data relatively. This can be a reason why the speed difference is diminishing when the ship speed is increasing.

Because of a large difference of ship speed, it is not easy to compare the power savings except maximum speed obtained from Sakura which is 15.35kts as shown in Figure 9. The power saving at this ship speed can be concluded as 17-18%.

By analysing the trial data, wake factor and resistance curves at towing condition can be obtained.

If these figures are close to the prediction, this also guarantee the accuracy.

$Kq = DHP / 2\rho \pi n^3 D^5$

$J = Vs (1-w) / nD$

$\eta_D = (1-t)/(1-w) * \eta o(J) * \eta_R$

$EHP = DHP * \eta_D$

This equation can be applied all the mean values of double run and the accuracy of measured ship speed were examined. It seems that predicted EHP values of Shigenobu is slightly lower than Sakura.

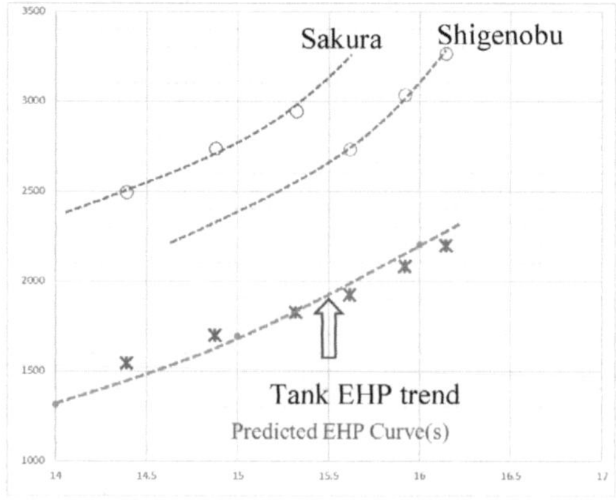


Figure 9 Predicted effective horsepower curves.

### 2.4.2 Other factors to be considered

The hull surface condition is the most important factor to guarantee accuracy of the measurement. The maintenance history of the two ships related this issue can be seen in Figure2 (bottom).

From this information, we concluded that the hull surface condition of Sakura will be better than Shigenobu because the full blasting and the totally new paint was applied to the Sakura on Jan. 2019 that is after the delivery of Shigenobu in Dec. 2017.

Another important factor is the effect of wind and waves on the performance of both vessels. The joint sea trial was planned to eliminate this effect however this assumption can be appropriate only if the ship speeds of both ships are close.

.

## 2.5 Zig Zag Test

The course keeping ability is the most important factor for the manoeuvrability of a container ship to avoid accidents and cargo loss during severe weather conditions because the rolling motion of a container ship is strongly affected by poor seakeeping ability.

Photo 3 A photo taken from the Sakura while she is following Shigenobu.

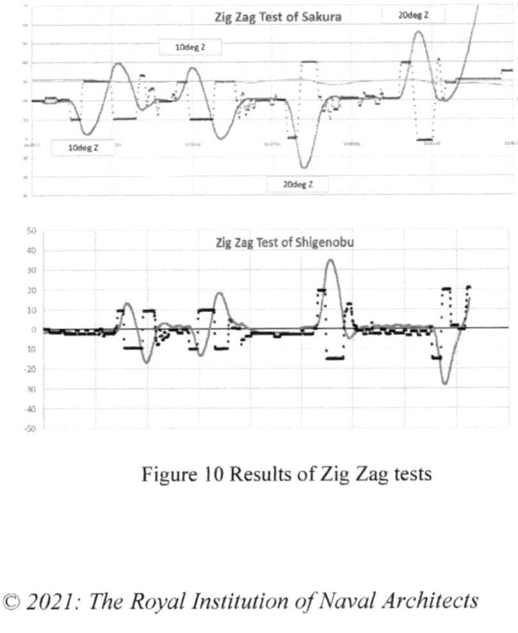


Figure 10 Results of Zig Zag tests

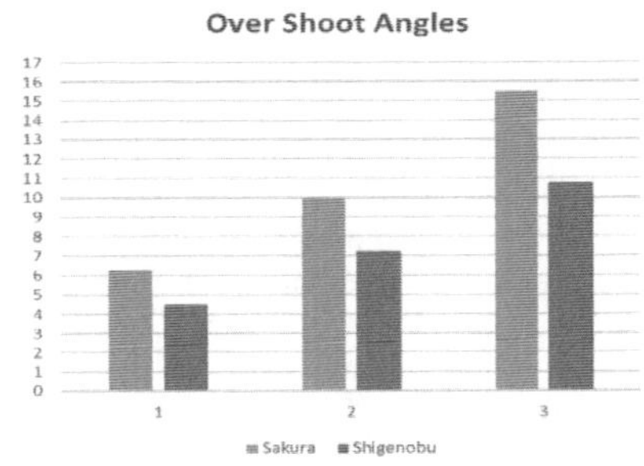


Figure 11 Overshoot angles gathered from the zig zag tests.

The test results can be seen in Figure 10 and it can be summarized as Figure 11. The overshoot angles of Shigenobu are clearly less. It is obvious that Gate Rudder ® has a great potential to improve a ship course keeping ability.

## 2.6 Performance of Berthing at narrow port

The crabbing mode is a strong steering function during the port operation which can increase the ship operability and reduce the risk of damage with hull and cargo. The thrust direction is almost lateral (80-85deg) and the stern thruster can be replaced by the Gate Rudder when a small advance speed is accepted.

The two ships, Sakura and Shigenobu, tried to berth the same quay where the distance from the other side quay is only 130m while the ship length is 109m.

Sakura managed to berth (head in) with anchors, bow thruster and the flap rudder as shown in Photo 4 (lower) which is the usual operation under this windy condition. On the contrary, Shigenobu entered the harbour and executed a pivot turn using the crabbing mode function as shown Figure 4 (upper).

Photo 4 Shigenobu (upper) and Sakura (lower)

## 3 ANOTHER EXCELLENCE

### 3.1 Speed loss due to rudder helm

It is well known fact that the high lift rudder is showing higher resistance during the steering conditions because of not only additional drag but also induced drag resulting from the higher lift force. However, the Gate Rudder can generate this lift force with different directions as shown in Figure 12 same as tacking of a yacht while the propeller will deflect the flow towards to the drag directions.

This difference of speed loss due to helm can be seen clearly during the turning period of speed trials as shown in Figure 13.

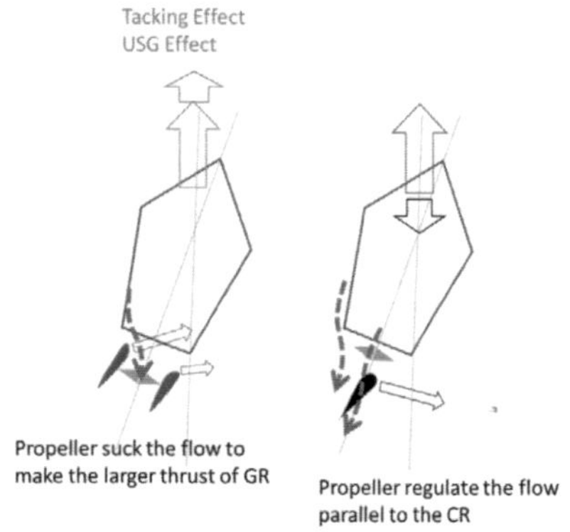


Figure 12 Tacking effect of gate rudder system

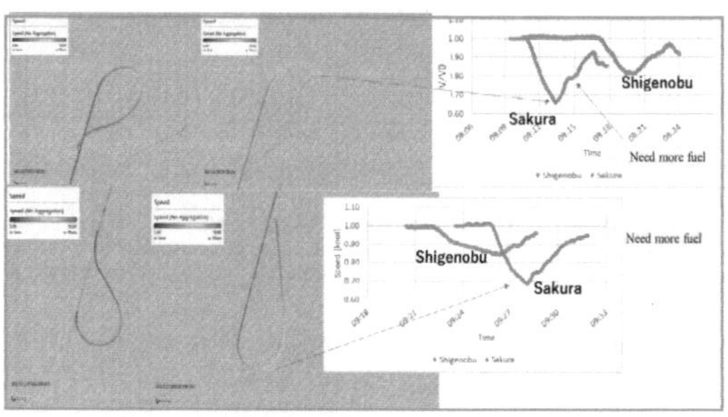


Figure 13 Speed loss due to turning motion.

### 3.2 Acceleration from zero speed

It is a well-known fact that the ducted propeller will perform well at bollard conditions. Therefore, we can expect that Shigenobu will show the better start-dash than Sakura under the same engine load (the same rpm).

The comparison was made at the beginning of this joint sea trial as shown in Figure 14.

Figure 15 also shows the remarkable difference of acceleration of both ships on a straight course indicated by the dotted area in Figure 14.

Table 4 shows the calculated acceleration at 1kts and 2kts for both Sakura and Shigenobu,

Here T is propeller thrust obtained from following equations:

$$(1+\alpha)\, M\, dV/dt = T\,(1\text{-}t) - R(V)$$

where M represents the mass of the vessel and $\alpha$ the factor for added mass.

This large difference (abt.1.8 times) of speed gradient requires the fact that gate rudder is producing 1.8 times larger thrust than Sakura's propeller-rudder system and thus one may say that the propulsive efficiency of the Gate rudder system during this acceleration phase is 1.8 times better than the conventional rudder system of Sakura..

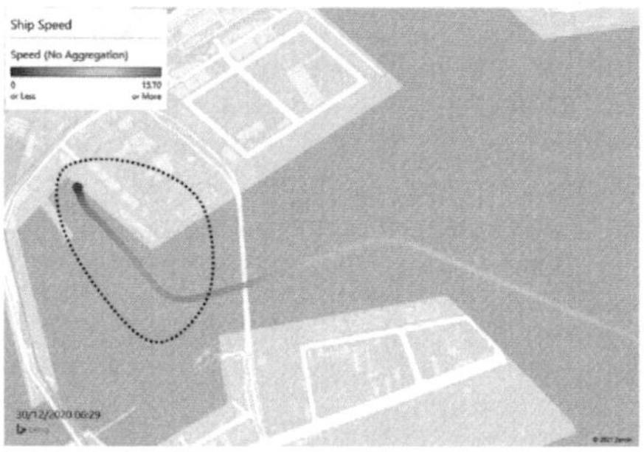


Figure 14 Start-dash point of two ships

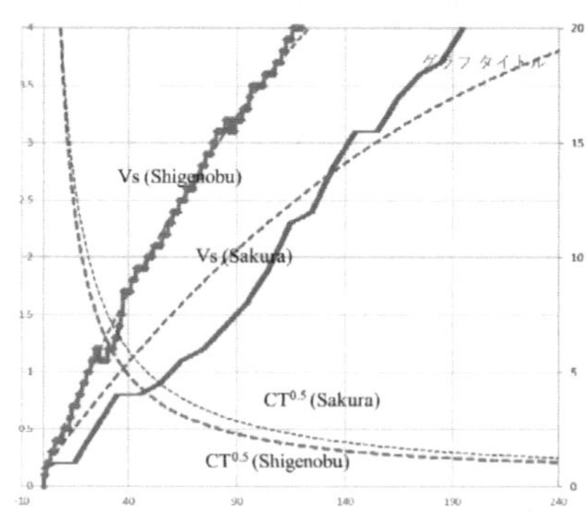


Figure 15 Trend of ship speed after zero speed.

Table 4 Obtained acceleration data from two ships

| | V(m/s) | dV/dt | T (kN) | $CT^{0.5}$ | ratio of T |
|---|---|---|---|---|---|
| Sakura | 1.02 | 0.012 | 52.93 | 4.02 | |
| Shigenobu | 1.02 | 0.020 | 98.12 | | 1.85 |
| | | | | | |
| Sakura | 0.51 | 0.015 | 62.03 | 9.22 | |
| Shigenobu | 0.51 | 0.027 | 109.78 | | 1.77 |

 

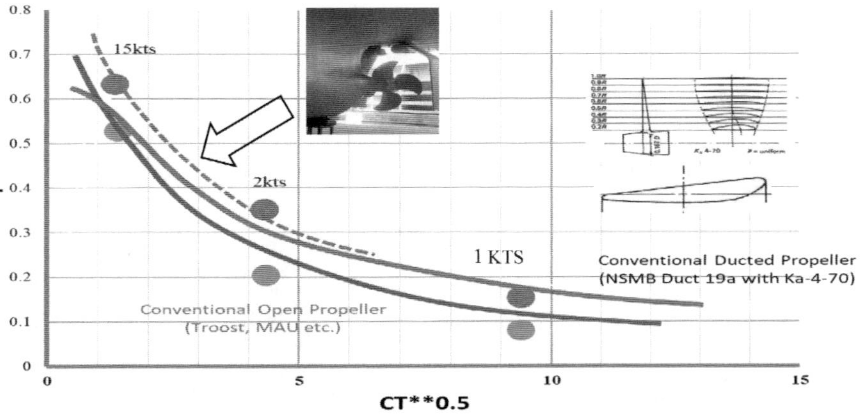


Figure 16 Gate Rudder Efficiency behind the ship compared with other conventional propellers

To confirm the mechanism, the optimum efficiency of Gate Rudder, the envelope of maximum efficiency of ducted propeller and conventional propeller was investigated.

If we select the point corresponding to 1kts (0.51m/s), the $Ct^{0.5}$ values will be 9.22 according to the Table 4, and we need 1.85 times larger thrust to obtain this difference between conventional propeller and gate rudder system. At the point 2kts, the difference of gradient is similar to the 1kts case. Figure 16 shows the estimated propeller efficiency of Sakura based on its propeller characteristics by blue dots, while the red shows the calculated points by multiplying the ratio of propulsion thrust of each rudder system which is 1.77 (1kts), 1.85(2kts) and 1.17 (15.35kts). These three red dots can explain well why the gate rudder shows so remarkable efficiency gain when we compared with conventional ducted propeller efficiency.

3.3 Ship Motion in wind and waves

The ship yaw motion of Sakura and Shigenobu was measured by installed VOG with 0.2 sec intervals and the rolling by readings by clinometer of both bridges. As shown in Figure 17, two vessels were running the same track with several hundreds meter distance. The wind and waves were coming around 110-120deg (port side). The measured amplitudes and frequency are summarized in Table 5.

Table 5 Ship motion on the same track

| | roll | yaw | Freq. |
|---|---|---|---|
| Sakura | 3-5deg | 0.6-1.0 deg | Abt.30sec |
| Shigenobu | 1-3deg | 0.3-0.5 deg | Abt.30sec |

Figure 17 Damping effect of gate rudder on rolling and yawing motion

## 4 CFD STUDY

### 4.1 Full-scale CFD approach

Parallel to the full-scale measurements on the two vessels, also Computational Fluid Dynamics (CFD) simulations have been executed to identify the differences in performance due to the Gate Rudder. Based on previous experiences, it has been decided to include the detailed propeller geometry and to use the time-dependent sliding interface solution method (see figure 18).

Simulations in straight sailing condition have been made with VOF approach for the free surface and free sinkage and trim of the vessel. Mesh cell count is in the order of 10-15 million. The final time-step size is set equivalent to 2° of a propeller revolution, in line with internal Wärtsilä guidelines [12-13]. Straight sailing CFD results showed fair agreement with sea trial data.

Figure 18 Geometry and surface mesh of aft-ship, propeller and rudder as used in CFD simulations

### 4.2 Turning circle simulations

Next to the more conventional straight-sailing speed-power CFD simulations, a new area has been explored on the CFD simulations related to vessel manoeuvring. Already in the 1960's the concept of determining the vessel manoeuvring coefficients with a rotating arm approach was considered. From experimental point of view this resulted in a rather large challenge to get a basin of sufficient large size. This problem has been circumvented with more complex measurement procedures like the pure-sway and pure-yaw motions, from which the different vessel manoeuvring coefficients can be derived.

When using the CFD simulations the basin-size is often not a critical issue. The option of coupled rigid-body motions in Simcenter Star-CCM+ allows for a relatively simple set-up of the vessel domain which rotates around a centre-point with given radius, with superimposed to this the vessel drift and the propeller rotation. An example of the set-up with its fairly small numerical domain is shown in figure 19.

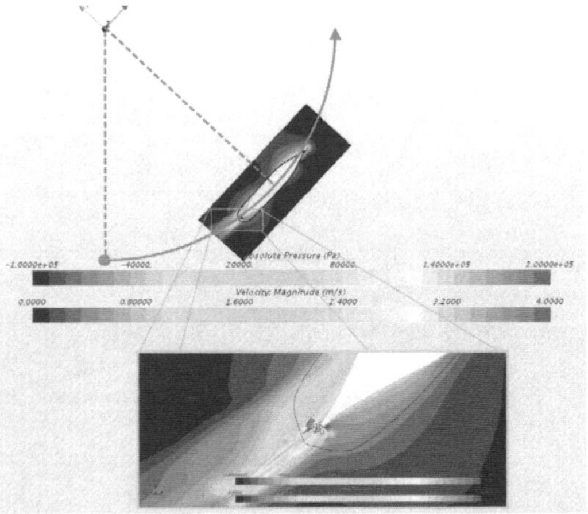

Figure 19 example output of CFD simulation based on rotating arm approach

As a result, a numerical flow simulation model can be made, where the vessel sails at prescribed turning circle diameter, with prescribed rudder angle and vessel drift angle. Outcome of the simulations give the total of the surge and sway forces and the yaw moment acting on the vessel, including the propeller and rudder and its interaction effects. The actual steady state turning circle condition can be found from interpolation, where the different surge and sway forces should be in equilibrium, as well as the vessel turning rate (yaw-moment).

From a large range of conditions one example case will be discussed in more detail, which is the turning circle of 4.5 D/L for a steady speed of 10 knots. The corresponding turning rate $r$=0.2 rad/s or in dimensionless representation $r'$=0.4. The required vessel sway force becomes 688kN for this condition to reach equilibrium.

The simulations have been executed for both the Gate Rudder and Flap-rudder configuration. First the vessel rudder angle and drift angle have been tuned to reach equilibrium on sway and yaw, and second the propeller RPM has been tuned to find the thrust-resistance balance.

Once the equilibrium of the two forces and the vessel moment has been determined, the required shaft power can be calculated from the shaft torque and RPM. Difference in performance can be seen clearly in figure 20, where the power levels of both Flap-rudder and Gate-Rudder are shown.

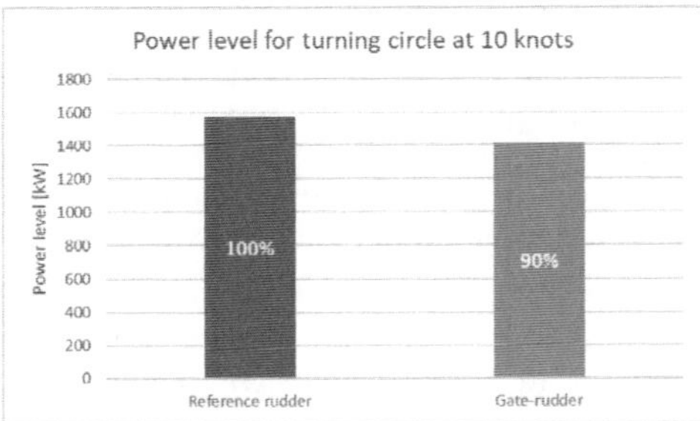


Figure 20 comparison of shaft power requirement to execute steady turning circle motion at 10 knots with D/L=4.5

### 4.3 Straight sailing with yaw

After the successful implementation of the rotating arm set-up in the CFD simulations to determine performance in a turning circle, a second phenomenon has been studied: straight sailing with yaw angle. This represents for example a steady drift due to an external side force, like the wind. In order to keep numerical consistency with the previously discussed simulations, the straight course has been mimicked by a rotating arm radius of 10km. With a set of combinations of rudder angle and vessel drift angle again the vessel yaw-moment equilibrium has been found for the straight sailing condition. The magnitude of the external side force has been chosen, which results in an actual vessel drift angle and rudder angle. The CFD simulations have shown that the vessel with Gate Rudder can operate at smaller drift angles compared to the sister-vessel with Flap-rudder. The results for a range of conditions are shown in figure 21. This trend is in line with the findings in the sea trials, shown in table 5. Besides the obvious variation of the rudder angle, also variations in vessel speed are included. The straight lines indicate that the slope of the line depends on the actual vessel and rudder geometry/type.

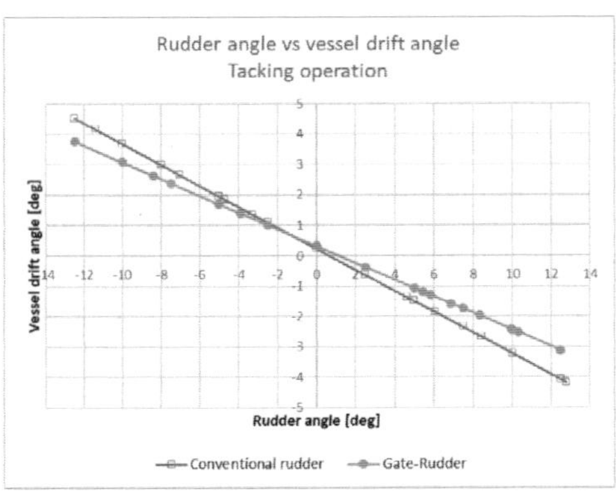


Figure 21 vessel rudder angle vs drift angle for straight sailing with yaw angle (tacking operation)

### 4.4 Vessel acceleration simulations

All above described CFD simulations address a steady state aspect of the vessel performance, either (1) straight sailing, (2) sailing in a turning circle or (3) sailing with steady yaw-angle. In these conditions the impact of vessel inertia is limited or even non-existing.

Evaluation of vessel acceleration on the other hand is governed by a large portion by the inertia effects. In order to simulate accelerating vessel performance, the scope of simulations is extended to system simulations [13]. The dynamic system simulations in Simcenter Amesim take the propeller performance under varying loads, the vessel inertia and drive-line limitations into account.

The simulated vessel acceleration is compared with the available sea trial data for the Gate Rudder in figure 22. The drop in vessel speed at time 06:10 is related to the sharp turn of the vessel in port as shown in figure 14. Agreement between the sea trial data and the dynamic simulations is reasonable. The simulation model predicts slightly faster acceleration in the time from 06:12 till 06:18. This could be related to the engine torque limitation, which is currently not implemented in the system simulation. The simulation model therefore allows for continuous ramp-up of the engine RPM, whereas the actual ramp-up in the sea trials might have been impacted by the torque limiter. This effect might be taken in future research on the Gate Rudder performance into account. It should also be noted that vessel acceleration simulations of the Flap-rudder case are currently not yet available due to the relative short time between the execution of the sea trials and the preparation of this paper.

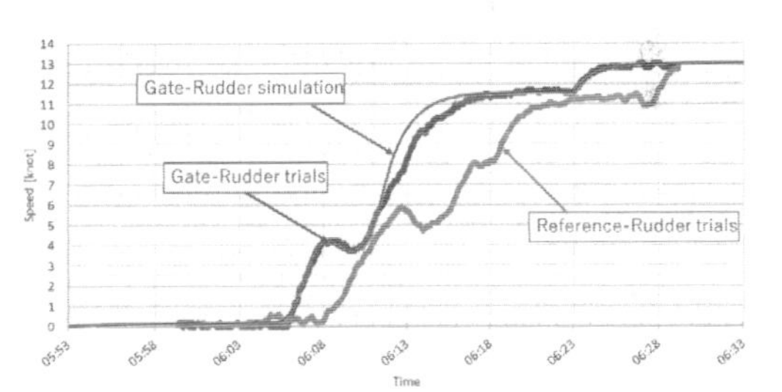


Figure 22 comparison of Gate-rudder vessel acceleration sea trial data and system simulation results

## 5 CONCLUSION

The full scale performance results obtained from a joint sea trial of two sister ships and full scale CFD study for acceleration phase and turning motion was presented.

The obtained results can be summarized as follows:

(1) Energy saving of 17-18% by Gate Rudder was confirmed by speed trials under the same weather and current conditions. The figures are very close to 19% that was obtained from voyage data of two ships on the same western route.

(2) The better course keeping ability of Gate Rudder was confirmed by zig-zag tests. This is important characteristics for the coastal ships from the safer operation point of view during the navigation among archipelago where strong current exists.

(3) The effectiveness of the crabbing mode was confirmed during the berthing operation. It implies that the stern thruster operation can be replaced by Gate Rudder because of its strong lateral force produced by overlapping rudder blades.

(4) Ship motion such as rolling and yawing can be remarkably reduced by Gate Rudder. The comparison was made on the same track where strong wind and waves were observed.

(5) .It was confirmed that the Gate Rudder acts similar as a ducted propeller at high propeller loading condition such as start-dash period. Moreover, it shows much better performance at normal propeller loading condition where the conventional ducted propeller loses part of its advantage.

(6) CFD was conducted to explain the mechanism of Gate Rudder. Both full scale data and CFD data show the excellent performance of Gate Rudder during the turning motion. Also tacking performance is in favour of the Gate Rudder.

(7) Full scale CFD simulations are a useful tool for designing a Gate Rudder and evaluating its performance

It seems that the Gate Rudder can be one of the best candidates which can be applied to solve the conflict between EEDI phase 3 and the minimum power requirement rules while further studies will be still needed.

## 6 ACKNOWLEDEMENT

The joint sea trial is the most difficult full scale test which affect on the commercial activity of a ship owner. We are deeply grateful to Mr. T. Imoto president of Imoto Line and his staffs who kindly supported this valuable joint sea trial.

We would like to express our gratitude to late N. Asaumi and S. Asaumi president of Yamanaka shipyard and his staffs who gave us the chance to conduct this test by providing all the information we need. As we introduced in the summary, the best hull form with low resistance of Sakura and Shigenobu was designed by Yamanaka shipyard.

Finally, we would like to express our special thanks to Tokyo Keiki Co. for their kind support with accurate measurements.

## 7 REFERENCES

1. Sasaki N, Atlar M., Kuribayashi S., Advantages of twin rudder system with asymmetric wing section aside a propeller, *JMST* 2015
2. Sasaki N. et.al., Measurement and calculation of Gate Rudder Performance, AMT2017
3. Kuribayashi S. et.al., Advantages of Twin Rudder System with Asymmetric Wing Section aside a Propeller (6th report) *JASMAOE*, Hiroshima , 2017
4. Kuribayashi S. et.al., Advantages of Twin Rudder System with Asymmetric Wing Section aside a Propeller (7th report) *JASMAOE*, 2018
5. Yilmaz N. et.al., Tip vortex cavitation simulation of a propeller in a Gate Rudder® system, *AYO Colloquium* 2018
6. Sasaki N., Atlar M., Investigation into the propulsive efficiency characteristics of a ship with the Gate Rudder propulsion, *3rd IMPPC* 2018
7. Sasaki N. Atlar M., GATE RUDDER, *INT-NAM* 2018
8. Fukazawa M. Turkmen S., Marino A., Sasaki N., Full-scale gate rudder performance obtained from voyage data, *AYO Colloquium* 2018
9. Turkmen S., Fukazawa M., Sasaki N., Atlar M., Cavitation tunnel tests and full-scale review of the first Gate Rudder system installed on the 400TEU container ship, *AYO colloquium* 2018
10. Sasaki N. Kuribayashi S., Miles A., Scale Effect of Gate Rudder and a New Power Prediction Method, *RINA Propellers & Impellers* ,2019
11. Sasaki N., Atlar M., Towards a Realistic Estimation of the Powering Performance of a ship with a Gate Rudder System, *SMP'19* 2019
12. BULTEN, N., 'With numerical simulations to more efficient ship designs', *Proceedings RINA Energy Efficient Ships conference*, London, UK, 2016
13. BULTEN, N., 'Usage of digital tools for the optimum propeller design', *Proceedings RINA Propellers & Impellers*, London, UK, 2019

## 8 AUTHORS BIOGRAPHY

**N. Sasaki** holds the current position of visiting professor at University of Strathclyde and a technical advisor of Kuribayashi steam ship Co. involved in several energy saving projects. He was working for Newcastle University as a visiting professor, MTI (Subsidiary of NYK), NMRI (National Maritime Research Institute) and SHI (Sumitomo Heavy Industries).

**S. Kuribayashi** holds the current position of the chairman of Kuribayashi stem Co. ltd. and the main patent holder of gate rudder. His innovative idea is implemented to the idea of gate rudder design and he contributed to installation of the first gate rudder on Shigenobu.

**N. Bulten** holds the current position of Product Performance Manager at Wärtsilä – Business line Propulsion. He is responsible for hydrodynamics within Propulsion business. He has more than 20 years of experience in the field of CFD and marine propulsion systems. In his current assignment he focusses on development and implementation of system simulations and advanced CFD-methodologies to include various Energy Saving Devices into the performance predictions.

**M. Yazawa** holds the current position of R&D group at Kamome Propeller Co., Ltd. He is responsible for designing propellers and CFD. He has contributed to the scheduling of the trial and analysing the result of sea trial.

**The Royal Institution of Naval Architects**
8-9 Northumberland Street
London, WC2N 5DA
United Kingdom

ISBN 978-1-7138-8297-8
9 781713 882978

CURRAN ASSOCIATES INC.
proceedings
.com

# Pedestrians 2018

Transportation Research Record: Journal of the Transportation Research Board

Volume 2672, Issue 35

SAGE Publishing